THE NEXT
FIFTY YEARS
IN SPACE

THE NEXT FIFTY YEARS IN SPACE

PATRICK MOORE

With drawings by Andrew Farmer

Taplinger Publishing Company
New York

First published in the United States in 1976 by
Taplinger Publishing Co., Inc.
New York, New York

Reprinted 1978
Limpback edition first printed 1978

Printed in Hong Kong by
Mandarin Publishers Ltd.

Library of Congress Catalog Card Number: 75–26326
Cased ISBN 0-8008-5528-0
Limpback ISBN 0-8008-5529-9

CONTENTS

ILLUSTRATIONS

To Barney D'Abbs
Who sees the future as I see it — and whose
glimpses ahead have helped my own!

FOREWORD

Making forecasts is always dangerous, particularly in a fast-developing subject such as astronautics. Yet there is no harm in speculating, and this is what I have tried to do. Only time will tell whether I am right or whether I am grotesquely wrong!

I must express my most grateful thanks to my friend of many years, William Luscombe, without whose encouragement the book would not have been written, and who must surely be an author's ideal publisher.

Selsey,
Sussex

PATRICK MOORE

1
PROLOGUE:
THE STONE AGE
TO 1975

On the night of 20–21 July, 1969, I was broadcasting "live" from the BBC television studio in London. It was a great moment; *Eagle*, the lunar module of Apollo 11, was making its final descent toward the rocks of the Moon. The tension even in our studio was something never to be forgotten, while at Mission Control at Houston, Texas, it must have been unbearable. Then, at last, we heard Neil Armstrong's voice coming to us from a distance of a quarter of a million miles: "The *Eagle* has landed."

It was the start of a new era. Another world had been reached; Man's isolation was ended, and nothing could ever be quite the same again. Yet despite the magnitude of the triumph, it was not unexpected. For some years it had been obvious that unless any major setback occurred, the first lunar voyage was imminent. Things had been very different not so long before; and in beginning a book which must, inevitably, be both personal and speculative, I think I may be allowed to begin by recalling a few of the encouraging forecasts made earlier by highly eminent people. Here, then, are a few typical specimens:–

Dr. Dionysius Lardner, addressing the British Association in 1838. – "Men might as well project a voyage to the Moon as attempt to employ steam navigation across the stormy North Atlantic Ocean."

The New York *Times*, in a 1920 comment upon Professor Robert Goddard's suggestion of sending a charge of flash-powder to the Moon. – "A Severe Strain on Credulity . . ."

Professor A. W. Bickerton, of Auckland University, one-time tutor to the great atomic scientist Rutherford. – "This foolish idea of shooting at the Moon, revived both in America and in Moscow, is an example of the absurd lengths to which vicious specialization will carry scientists working in thought-tight compartments."

Professor Forest Ray Moulton, one of the most distinguished astronomers in the United States, writing about space-travel in 1935. – "It must be stated that there is not the slightest possibility of such a journey. . . . There is no theory that would guide us through interplanetary space to another world even if we could control our departure from the Earth; and there is no known way of easing our ether ship down on the surface of another world, if we could get there."

Editorial comment in the London *Daily Mirror*, 18 October 1948: "Our candid opinion is that all talk of going to the Moon, and all talk of signals from the Moon, is sheer balderdash – in fact, just moonshine."

Of course, any new venture meets with criticism of this kind. Aeronautics is a case in point. Even after the Wright Brothers had made their pioneer flights from Kitty Hawk, Professor Simon Newcomb – like Moulton, an eminent American astronomer – was still proving, to his complete satisfaction, that the only way to make a heavier-than-air vehicle fly was to tow it by large numbers of little birds. Remember, too, that in 1934 the British Under-Secretary of State for Air wrote a famous letter to the effect that the method of jet propulsion could never be a serious rival to the airscrew-engine combination, so that the Government did not propose to spend any time or money in studying it. I have no doubt that in the Stone Age similar comments were made about the invention of the wheel. But in the case of space-travel, official scepticism lasted for a surprisingly long time, and was not finally dispelled until 4 October 1957, when the Russians sent up the pioneer artificial satellite, Sputnik 1.*

Even before the war, when I was in my teens, I was convinced that flights to the Moon would be made in the foreseeable future, but my time-scale was wrong. I expected the first lunar landing to take place some time between 1990 and 2010, so that my estimate was too pessimistic by more than twenty years, though I plead that most people were even less accurate – and only a few, notably

*Most people remember the comment by Sir Richard Woolley in January 1956: "Utter bilge." In fact, this was an off-the-record remark made when Sir Richard had just completed a long and tiring air journey, and found himself surrounded by clamouring journalists. During his spell as Astronomer Royal he played a distinguished rôle in many space research committees, and to put any emphasis upon a single tired comment would be most unfair.

Arthur Clarke, were close to the truth. The progress of astronautics was speeded-up tremendously by the war; we have to admit that the German V.2 weapons were the true ancestors of the Apollo vehicles, even though they were designed for quite different reasons and were, by modern standards, primitive. All the same, man would probably have reached the Moon before A.D. 2000 even if Hitler and Mussolini had never existed.

What I propose to do, in the present book, is to look ahead and see what is likely to happen during the next fifty years – that is to say before A.D. 2025. I am well aware that my forecasts may be very wide of the mark, and if any copies of the book survive in 2025 they will be read with caustic amusement. I must, of course, stress that the views given here are entirely my own, and others will not agree; but I will do my best.

As a prologue, it seems necessary to say something about past history, since otherwise it will be difficult to see things in their proper perspective. And let it be made clear at once that though practical astronautics did not begin until well into our own century, the idea of travelling to the Moon is very old indeed. So far as I know, the first of all science-fiction stories was written by a Greek, Lucian of Samosata, around A.D. 150, at a time when the Earth was still supposed to be lying at rest in the centre of the universe with the entire sky revolving round it. Lucian's book was called the *True History*, because, as he pointed out, it was made up of nothing but lies from beginning to end. He described how a ship making its way through the Pillars of Hercules (in modern geography, the Straits of Gibraltar) was caught up in a waterspout, and hurled aloft so violently that after seven days and seven nights it landed on the Moon. On arrival, the sailors found that the King of the Moon was about to wage war against the King of the Sun, because both monarchs claimed sole rights with regard to Venus (the planet, that is to say, not the goddess). Later writers were equally ingenious, if no more practical. In 1634 Johannes Kepler's hero travelled to the Moon by demon power, and four years later came a famous novel by Bishop Godwin, in which the astronaut was taken moonward on a raft towed by wild swans. Meantime, if legend is to be believed, the first actual experiment had taken place. It is said that a Chinese nobleman named Wan-Hoo built a framework equipped with forty-seven rockets, and a saddle in the middle. When he sat on the saddle, and ordered his retainers to light all forty-seven rockets at once, the results were exactly as might have been predicted.

Actually, the first serious attempt to put forward a workable method was made by no less a person than the great French writer Jules Verne, in 1865. Verne was not himself a scientist, but he believed in making his stories as authentic as possible, and *From the Earth to the Moon* was no exception. (The sequel, *Round the Moon*, appeared a few years later.) His heroes were shot to the Moon in a hollow projectile, fired from the mouth of a huge cannon built specially for the purpose. They left at a velocity of 7 miles per second, or about 25,000 m.p.h.; they made a circuit of the Moon, and then fell back to their home planet, to be fished up out of the sea quite unharmed.

Verne's story is a museum-piece, but it is still worth reading, and there is much to be learned from it. His proposed launching site was in Florida, not too far from the present-day rocket base at Cape Canaveral; he described a giant telescope on Long's Peak which might be taken as a forerunner of the Palomar 200-inch reflector; and the first real lunar travellers, the crew of Apollo 8, were indeed picked up after landing in the ocean. Where Verne went wrong was in his method of propulsion, and in his ideas about zero gravity or weightlessness. I must dwell briefly upon these, because they are so vital in any discussion of flight beyond the Earth.

He was basically right in saying that the departure would have to be made at 7 miles per second, because this is the Earth's escape velocity. Starting off at a lower speed would mean that the projectile would be unable to break free from the pull of gravity, and would fall back before going anywhere near the Moon. Unfortunately, the shock of blasting off at 7 miles per second would promptly turn the luckless travellers into jelly; and in any case the friction set up by moving through the dense bottom part of the atmosphere at such a rate would burn the projectile up immediately. It is not impossible that the space-gun principle may have its uses in the future, but it will be confined to firing non-fragile payloads off airless worlds, and so far as the Earth is concerned it is out of the question. Moreover, the journey would be a one-way trip only. Verne evaded this difficulty by introducing a minor satellite of the Earth, which swung his projectile out of its planned path and brought it home; but there is no such satellite — and even if there were, it would not act in such a manner. All in all, space-guns built on Earth must be firmly relegated to the realm of science fiction.

Next, there is the question of zero gravity. Verne believed that his travellers would become weightless only when they reached a "neutral point" in space, where the Earth's gravitational pull was exactly balanced by that of the Moon. In fact, they would have been weightless from the moment of departure, and the so-called neutral point is of no importance whatsoever in this respect.

Of course, there is no need to start off at 7 miles per second if you can go on using power throughout the journey, but the space-gun allows for a single initial burst and then nothing more. Neither is there any point in considering vehicles such as balloons or aircraft, because all these are useless except when surrounded by atmosphere; and above a few tens of miles the Earth's air becomes so thin that to all intents and purposes we can forget about it. Most of the quarter-million mile journey to the Moon has to be done in "empty" space, and this involves some radical re-thinking.

Various suggestions were made during the nineteenth century. One of the early theorists was a Russian named Kibaltchitch, whose career was cut short because he was unwise enough to provide a bomb used to assassinate the Tsar of Russia, and the authorities not unnaturally took grave exception to anything of the sort. Then there was Hermann Ganswindt, an eccentric German who spent his life in designing weird and wonderful machines which signally failed to work; I need mention only his helicopter, which was of limited value because he could not provide it with an engine. His idea of a spaceship was to have a cylindrical plate, below which charges of explosive were set off in succession; each time the plate banged against the roof of the vehicle, the whole contraption was propelled upward. But the real break-through was made by Konstantin Tsiolkovskii, a shy, deaf Russian schoolteacher who worked out the principles of space-flight as early as 1895, and whom the Russians justifiably regard as "the father of astronautics".

Tsiolkovskii realized that the only possible method of going into space was by making use of what is called the principle of reaction. Consider a firework rocket, of the type let off on Guy Fawkes' Night. Here we have a hollow tube, filled with gunpowder; when you "light the blue touch-paper and retire immediately" the powder burns, and produces hot gas; this gas rushes out of the tube through the exhaust, and in so doing it propels the rocket body in the opposite direction. In fact the rocket is pushing against itself, and there is no need for any surrounding air. Actually, air is a handicap, because it sets up friction and has to be pushed out of the way. The rocket works at its best when moving in vacuum.

Tsiolkovskii also saw that solid fuels of the gunpowder variety are too weak and too uncontrollable to send a vehicle to the Moon, and instead he proposed to use liquids. The principle here is to mix two different liquids in a combustion chamber, so that they react together and produce hot gas which is expelled from the

17

exhaust just as with the Guy Fawkes firework. The thrust available is much greater, and the simple powder pack is replaced by a genuine rocket motor. Yet it is still not strong enough to reach escape velocity, and Tsiolkovskii's remedy was to build a compound vehicle made up of several rockets mounted one on top of the other. The massive lower stage would use up its propellant, and would then break away and fall back to the ground, leaving the upper stage to continue the journey by using its own engines. In fact, the upper stage is given a running jump into space, and theoretically it is possible to use any number of stages.

This is essentially how modern space-launchers work, and there can be no doubt at all that Tsiolkovskii's ideas were in many ways far in advance of their time. Yet he was purely a theorist; he never built a rocket in his life, and was in no position to do so. Moreover, his papers were published in obscure Russian journals, and caused about as much impact on scientific thought as a feather falling upon damp blotting-paper. His first important contributions were written in 1903, but it was not until shortly before his death, in 1934, that he became famous in the Soviet Union. By then the first liquid-propellant vehicles had been launched, and the main credit must go to Robert Hutchings Goddard, of the United States.

Goddard had never heard of Tsiolkovskii, but in 1926 he fired a modest vehicle which reached a speed of 60 m.p.h., and travelled for a total distance of almost two hundred feet. Others followed. Not much was heard about them, because Goddard was no lover of publicity; but things were also moving in Germany, largely because of the interest sparked off by a book by one Hermann Oberth. Oberth, who was born in Roumania, is still alive, and it is pleasant to record that he was present at Mission Control, Houston, to watch the first lunar landing so many years later.

The German amateur rocket group included some famous pioneers, among them Wernher von Braun, whose rôle in the development of astronautics is unique. Here again the first successes were independent – the Germans knew nothing about Goddard's work, just as Goddard had known nothing about Tsiolkovskii – but they were promising. There were, of course, some bizarre episodes, and one test was actually financed by the City Council of Magdeburg in order to test a weird idea that the Earth's surface is the inside of a hollow sphere, so that a rocket fired vertically upward to a sufficient height will land in the Antipodes. (To be precise, two tests were made, but the first vehicle rose to a mere ten feet, while the second blasted off sideways instead of upward. I may add that von Braun and his colleagues had no faith in the hollow-globe idea, but any money provided for research was grist to the mill!) Alas, preparations for the coming war intervened. The Nazi Government took over all rocket research lock, stock and barrel. Some of the pioneers left hurriedly for more peaceful countries, while others, including von Braun, were installed at Peenemünde, an island in the Baltic, with instructions to forget about reaching the Moon and concentrate upon producing a rocket weapon which would prove decisive in Hitler's plans to conquer the world.

Undoubtedly it was the Nazi phase which gave the rocket its evil reputation. There had been earlier war-rockets; Britain had used them in the Napoleonic wars with a certain amount of success, but these weapons were of the uncontrollable, solid-fuel variety, and had been given up because they could not compete with more conventional types of armament. Peenemünde was something quite different. Fitted with powerful warheads, high-flying rockets would be very difficult to combat, as some of the members of the German High Command realized. By 1942, with the war at its most crucial phase, von Braun's team made the first successful test of the vehicle which was originally called the A4, but which became notorious as the V2. As I have said, it was the true ancestor of the modern space-ship, though it did not achieve escape velocity and was not intended to do so. It was a liquid-propellant rocket of a new order of magnitude; compared with it, all earlier vehicles were mere toys. Had the V2 become operational earlier than it did, the whole course of the war might have been changed. Fortunately Hitler's faith in it varied, and the first onslaught was delayed until 1944. It did not come as a complete surprise, and indeed Peenemünde had been bombed by the R.A.F. a year earlier; from my own point of view it is ironical that I might easily have taken part in that raid, though in fact I was flying elsewhere on that particular night. Between then and the virtual collapse of Germany the V2s caused considerable damage and casualties in the London area, and thoughts of reaching the Moon were very far from all our minds.

Peenemünde itself was eventually overrun by the advancing Red Army, but von Braun and most of the other leading scientists had already left, and had surrendered to the Allies. Before long they were in America, still working on rocketry, but now with a different aim. The United States Government had become rocket-minded, and a proving ground was set up at White Sands in New Mexico; undamaged, captured V2s were used as test vehicles, and in 1949 the first step-rocket achieved a record height of almost 250 miles.

It had become obvious that the rocket was destined to be an all-important scientific tool as well as a military weapon.

Next came the establishment of the rocket site at Cape Canaveral. (Subsequently the name was changed to Cape Kennedy, and then back again.) Testing went on apace, and one of the most significant decisions in world history was made in 1955, when the White House announced its intention of launching an artificial earth satellite.

The main drawback of the scientific rocket had always been that it could be used only once, and could operate at high altitude for only a very limited period. When it crashed back to the ground, it not only destroyed itself, but very often destroyed its instruments and records as well, though naturally every effort was made to detach the payload and bring it down gently by parachute. There were also some additional hazards, because not every rocket behaved according to plan, and there were even a few incidents which caused raised eyebrows in political circles. (One American missile crash-landed in – of all places – Cuba, and killed a cow, which was promptly given a State funeral as a victim of Imperialist aggression.) Something less wasteful than an ordinary rocket was needed, and the only logical answer was a satellite.

"Very well," I can imagine some people saying. "Launch a satellite by all means, but why will it not come down again?" The answer is: "Because it will be moving." The unusual analogy here is that of a stone being whirled around on the end of a string; so long as the string is kept taut, the stone will keep on circling your hand – and the force of gravity on a moving satellite may be compared with the action of the string on the stone. In fact the analogy is not good, and I repeat it here only because I have never been able to think of anything better. It is probably more useful to consider the Moon, which lies at approximately 240,000 miles from us, and takes just over 27 days to complete its circuit. It never swings in on a collision course with the Earth, because of its motion. There is nothing to brake it, and it simply "keeps on keeping on". An artificial body, taken out beyond the Earth's resisting atmosphere and given the right initial velocity, will behave in exactly the same way.

The American plan, announced from the White House on 29 July 1955, was to send up a football-sized vehicle in a rocket, and put it into a path which would make it circle the Earth in a period of rather less than two hours. The launching, it was said, could take place during the period between mid-1957 and the end of 1958 – the time of the so-called International Geophysical Year or I.G.Y., when scientists from almost all nations had agreed to co-operate in a programme to study all aspects of Earth science. Shortly afterwards two Soviet scientists, Sedov and Ogorodnikov, issued a statement to the effect that Russia was making similar preparations. Few people took much notice; very little news about space-research operations had come out of the U.S.S.R. since the end of the war.

When Sputnik 1 soared aloft, on 4 October 1957, the Western world was frankly taken aback. Not all the reactions were favourable, and a particularly illuminating comment was made by Rear-Admiral Rawson Bennett, Chief of Naval Operations in the United States, who said sourly that the satellite was "a hunk of iron that almost anyone could launch". However, at that time the Americans were quite unable to launch a hunk of iron or anything else. Their programme had run into trouble; inter-service rivalry had something to do with it, and it was not until the following year that von Braun was allowed to have his way and launch the first U.S. satellite, Explorer 1. By then the original Sputnik had been followed by another, which was far more massive and which carried a live dog, Laika.

Laika was the pioneer space-traveller, but she was doomed from the moment of blast-off, because there was no known way of bringing her or her rocket gently back to the ground. Today, when the so-called re-entry problem has been solved so completely, this may sound strange, but the situation then was very different, and all sorts of problems remained to be solved.*

The Soviet space programme, then, began in 1957; the American in 1958. I am writing these words in 1975, less than two decades later, and the progress which has been made is absolutely staggering. We have had scientific satellites, some of which have provided information which could never have been gained in any other way; we have had communications vehicles, so that by now there is full radio and television coverage all over the civilized world; men have remained in orbiting satellites for periods of up to nearly three months; the first true space-station has been set up;

*The launching of a dog to certain death caused immense controversy. Most people maintained that it was justifiable in the interests of science; certainly much was learned from the recording instruments strapped to Laika while she circled the world, and her end was probably quite painless. There were a few dissentients, including myself. I expressed my whole-hearted disapproval, and was criticized for so doing, but I am unrepentent. It may be added that we in Britain, where the law still allows stag-hunting, fox-hunting and similar barbarities, were really in no moral position to criticize the Russians.

astronauts have walked on the Moon; unmanned probes have been sent out to Venus, Mars, Jupiter and Mercury; and we have even launched one vehicle toward interstellar space, though whether any alien civilization will manage to salvage it seems rather dubious. If this has been accomplished in a mere eighteen years, what will the next eighteen years bring?

Oddly enough, I am not sure that the rate of progress will be quite so spectacular. Remember, this first hectic period was the time of pioneering. The rocket had suddenly come upon the scene as a powerful scientific tool, and one development led straight on to another. There is, I believe, something of a parallel with the beginning of what we may call modern astronomy. The telescope was invented around 1608, and in 1609–10 Galileo was able to use it to make one dramatic discovery after another. Then, inevitably, there was a lull which lasted for a long time. I doubt whether the same will be true of space research, but the analogy is there.

So let us see where we are likely to go next – beginning, as is only proper, with space activities close to our own home, this Earth from which we have at last managed to break free.

2
SPACE SATELLITES: 1975 TO 1990

To begin our look into the future, we must consider artificial satellites of all kinds – civil and military, manned and unmanned, and so on. The situation in the early part of 1975 was that many hundreds of satellites had been launched, mainly by Russia and America but also by a few other nations: Japan, China, Italy, France and Britain. It is sad to have to put Britain at the tail-end of the list, but we are still suffering from the after-effects of the restrictions imposed by the Explosives Act of 1875, which so effectively stopped the rocket pioneers here from carrying out practical experiments in the vital early period when von Braun and his colleagues were hard at work, first at their Rocket Flying Field near Berlin and then at Peenemünde.

Quite apart from their scientific value, the satellites have had tremendous effects upon everyday life, because they have revolutionized the whole of our communications systems as well as making important contributions to meteorology, Earth resource studies, and much else besides. There have also been some major space-station experiments. The Soviet Salyut 1 cannot be regarded as a success, inasmuch as its crew died during the return trip, but America's Skylab and later Russian Salyuts worked well. The casualty rate among astronauts has been astoundingly low, and there seems no obvious reason why things should not forge ahead rapidly. Alas, there are several "buts" which have to be taken into account not only with regard to satellites, but also affecting the whole of the rest of the space programme.

The worst of these is the danger of a nuclear war which would cause a fantastic amount of damage, and which might even bring civilization to an unpleasant and permanent end. The risk has been with us ever since the first atomic bomb fell on Hiroshima in 1945, and whether or not it has receded since then is a matter for debate. Previous wars have had no more than temporary and mainly local effects. Rome destroyed Carthage and drove a plough over its site; had Carthage destroyed Rome instead, Sputnik 1 would still have been launched, and you would still be reading this book. It is also true that the last war speeded up Man's progress into space, and at the crucial stage the German team could draw upon the whole resources of the Third Reich. But a major war today would have a very different result, and at best the devastation would take centuries to repair.

This is a political problem, though much too vital to be left to the judgement of politicians. I do not propose to discuss it at length here, because the ramifications are too great, but I must point out that the rocket is not the only scientific device to have been misused for military purposes; the same applies to the internal combustion engine, the airship, the aeroplane and practically everything else. Orville Wright (whom I once met) was saddened to reflect that his epic "hop" at Kitty Hawk had led on to the wholesale destruction of cities by fleets of bombers; but it was not his fault. I suggest that the same applies to most of the rocket pioneers, even those who, such as von Braun, spent years working upon vehicles which were quite openly meant to spread destruction.

The second consideration is finance. Space research is not cheap, and by everyday standards the amount of money involved is colossal. There have always been lobbyists who call for an end to the whole programme, so that the money saved can be diverted to "feed the starving millions", etc. On the face of it the demand looks reasonable: but is it? First, nobody but an idealist living in an ivory tower would expect the money saved to be put to good use, and it would almost certainly be used to buy more armament of the conventional type. Secondly, the amounts involved are very small when compared with a national budget. During one period the Americans spent 25,000 million dollars on the Apollo programme – and 150,000 million dollars on the war in Vietnam. During 1970 the British people spent seventy-five times as much money on alcohol as the Government did upon space projects.

Naturally, one must keep a sense of perspective; and if we could put the world to rights by cancelling all space research, then space research would have to wait. Yet things are not nearly so simple as this, and no thinking person will pretend that they are.

APOLLO 17
Apollo 17 on its launching pad, less than 24 hours before blast-off.
Photograph taken by Patrick Moore. Before launch, the gantry was of
course removed.

PATRICK MOORE AND APOLLO 17
The author with the Apollo 17 vehicle in the background, taken on
the day before launch.

FULL SCALE SPACE STATION (*overleaf*)

APOLLO 11: ALDRIN
The second man on the Moon: Colonel Edwin
Aldrin stepping down on to the lunar surface
from Apollo 11 on 21 July 1969. The photograph
was taken by Neil Armstrong, who had stepped
on to the Moon a few minutes earlier.

EVA 3
Astronaut Harrison H. Schmitt, at the foot of a
large boulder at North Massif in the Taurus–
Littrow area of the Moon: Apollo 17, the last
Moon mission to date. Dr. Schmitt was the first
qualified geologist to reach the Moon. The LRV
(Lunar Roving Vehicle) is seen in the foreground.

Thirdly, we must remember that we are entering a new environment, and there is always the danger that something will go disastrously wrong. Before Yuri Gagarin made his round-the-world flight in Vostok 1, in April 1961, nobody had any real idea of the possible effects of zero gravity, since this is something which cannot be simulated on Earth except for brief periods under very different conditions. It had been suggested that an astronaut (or a cosmonaut) might be promptly "space-sick", and would be in no state to carry out scientific observations. There were also misgivings about the various harmful radiations which come in from space, and from which we on the ground are shielded by the layers in our atmosphere. Another popular bogey was meteorite impact, and it had been maintained that any spacecraft moving above the thick part of the air would be promptly battered to pieces by a sort of cosmical bombardment. Gagarin's trip disproved all these pessimistic theories, and Skylab has shown that even after spending more than eighty days under zero gravity an astronaut can still walk unaided out of his capsule when he returns to terra firma. But it would be foolish to be over-confident, and we do not yet have all the answers.

Moreover, suppose that a space-station were built and manned – and then met with some catastrophe, with the loss of its entire crew? Or suppose that a manned venture to the Moon resulted in a crash-landing from which there could be no recovery? If anything of the kind happened, then space research would be held up for many years. Even if the causes of the tragedy were pinpointed, and precautions taken to prevent a recurrence, the anti-space lobbyists would seize their opportunity and make the most of it. Like everyone else, I fervently hope that there will be no such disaster, and the chances are definitely against it, so that I do not propose to take it into account in my time-scale. All the same, it is not impossible.

Actually, nothing is likely to halt the development of military satellites, for which Governments will always provide adequate funds. The old concept of "war in space" has long since been jettisoned, and the picture of interplanetary battle-craft locked in mortal combat belongs to the realm of science fiction. Equally absurd is the idea of building, say, a vast space-mirror which could focus the Sun's heat on to a hostile country and reduce its towns and forests to smouldering ashes. But reconnaissance is another matter, and here an orbiting satellite is vastly superior to any aircraft, because it can fly unrestricted over any part of the world. There is no doubt that American satellites have provided detailed information about Russian airfields, launching sites and installations of all kinds – or that the Kremlin has obtained similar information about the United States. It has been estimated, probably with truth, that over 70 per cent. of Russian and over 50 per cent. of American satellites are purely "military", using the term in a broad sense. This is a most depressing state of affairs.

An even more sinister concept is that of the Fractional Orbital Bombardment Satellite, or FOBS. A nuclear missile could be carried on board, and landed at the appropriate place and time; it would be extremely difficult to intercept, and of course it might arrive from almost any direction (whereas American strategy, for instance, is based on the assumption that any Russian attack would come from the north). Worse, a FOBS could be left in orbit to await a command from its controllers, and to destroy a satellite on the assumption that it might be a FOBS would spark off an international crisis more serious than any since 1945. Not having access to the thoughts of either the Pentagon or the Kremlin, I cannot say whether a FOBS has actually been tested; I hope not, but I have misgivings about it.

Nuclear weapons have not been used in conflict since Nagasaki because no one nation has the monopoly of them. The same applies to the "aggressive" military satellite, and we can only hope that it will never be tested in anger. All the same, I fear that the FOBS will remain in the minds of military strategists for some time yet – in fact, until the final end of the senseless Cold War and the beginning of a real understanding between the nations. When this will be, only time can show; but with any luck at all it should be before 2025.

Earth resources satellites are very different, even though they have inevitably some points in common with reconnaissance or spy-satellites. The first of them, Earth Resources Technology Satellite (ERTS) 1, was launched in July 1972, and was put into a polar orbit at a mean altitude of 500 miles. By using various photographic techniques, it could send back pictures showing which areas of crops were diseased and which were not; the value of this is obvious enough, as the saving in labour is enormous, and prompt action can be taken to deal with any infected areas. Of course, ERTS had many other tasks as well; and a photographic satellite can always give advance warning of, say, a potentially dangerous forest fire or the threat of a flood.

Between now and 1990 there will be many other Earth Resources satellites. A few dozen of them should by then be orbiting the world, and the coverage will be both continuous and complete. We must also bear in mind the value of a satellite in showing where to look for materials which we must have if our civilization is to survive: minerals, water, and – above all, perhaps – oil. I doubt whether anyone needs to be reminded about the oil crisis, which is becoming more and more menacing each year. Oil-producing areas are limited in extent, and tend to be concentrated; but satellites may well reveal others – and we will need them all, so that during the next decade or two there will be satellites launched specifically for this purpose.

The amount of oil available is not limitless. If Mankind continues to use it up at the present fantastic rate, then sooner or later (perhaps sooner rather than later) the supply will run out completely; and I would be the last to advocate that we should locate all possible oil-fields, by using satellites, and then work them out. But if there are major undiscovered oilfields, we need to know where they are. We will then be in a better position to decide how much we dare take from them.

I have already referred to the advance warnings given by satellites, which have saved many people from the fury of tropical storms. Nobody has yet any real idea of how to control a hurricane, and in fact in the present state of our knowledge it cannot be done; the next best thing is to have so complete a coverage of the Earth that no developing storm can pass unnoticed. I envisage a full-scale "hurricane patrol" by satellite which will be internationally organized, and which will keep watch for twenty-four hours a day. The cost of such a patrol would be more than offset by the advantages of it, and if it is not in action before 1990 I will be very surprised. This leads on, in turn, to a network of meteorological satellites which will give constant information about the state of the Earth's atmosphere, and will undoubtedly help to improve the currently uncertain science of weather forecasting. There is nothing revolutionary in this; the first meteorological satellites went up in the 1960s, and the main need now is for more of them, carrying the latest types of instrumentation.

One potentially encouraging feature of the Earth resources and weather satellites is that they should be international, because the data they will provide will benefit everybody. Why should not this be the same with communications vehicles? The first Trans-atlantic telegraph cable was laid between Ireland and Newfoundland almost exactly a hundred years before the ascent of America's first satellite, Explorer 1 of 1958, and the first radio message across the Atlantic was sent out by Marconi in 1901 – even though he had no real idea of why his experiment worked. But in 1901 the world was much less populated than it is now, and since then the pressure

CLARKE ORBITS

A satellite moving round the Earth in a period the same as that of the
Earth's rotation period, above the equator, will seem to remain
'stationary' in the sky, and is ideally suited to be a television and
radio relay. Satellites moving in orbits of this kind were suggested by
Arthur C. Clarke in a famous article published in 1945, and I
therefore suggest that they should be known as Clarke orbits.

on all available circuits and wavelengths has grown amazingly. In any case, cables and radio waves can deal with sound only; for visual communication there is no alternative to a satellite.

Telstar, the pioneer vehicle of July 1962, was a mere 34 inches in diameter, and weighed less than 200 pounds; but it showed that the scope was almost limitless. The "passive reflectors" such as the Echo vehicles (which were the brightest artificial satellites ever launched; it has been said that they were seen by more people than has been the case for any other man-made object!) have long since been superseded, and today we have satellites in "stationary" orbits, keeping pace with the Earth as it spins. This, incidentally, was originally a British concept, as it was first put forward by Arthur Clarke in a famous paper published in *Wireless World* in 1945.

Inevitably, Big Business has come into the picture, and there is as yet no chance that Mr. X, in London, will be able to "book a satellite" for a conversation with Mr. Y in New York or Tokyo. Neither do I expect that this will be possible as early as 1990, because a communications satellite is an expensive item; not only has it to be designed, made, equipped and launched, but it cannot be expected to operate indefinitely without maintenance. Telstar itself failed after a few months, and was permanently silent by February 1963. Ironically, the coup de grâce was administered by the trapped particles produced by an atomic bomb, exploded at an altitude of 250 miles by the U.S. Department of Defense – a criminally foolish experiment which should never have been allowed by any Government, and which we hope will never be repeated.

Because communications satellites are bound to become more and more important as time goes on, I suggest that they will be built on a really elaborate and massive scale, and that they will be periodically serviced by human crews who will make regular visits to them. At the moment this is not practicable, and if a satellite goes wrong there is nothing to be done about it; but by 1990 the space-shuttle principle, about which I will have more to say in Chapter 3, should have been developed to a point where going out to service a satellite will be no more difficult than going by aeroplane from London to Cape Town. Satellites set up at suitable distances above the resisting atmosphere will be in permanent, stable orbits – and if they have to be moved around for any reason, they can be made to do so by relatively weak rocket motors installed in them.

In view of all this, there should be a whole fleet of communications satellites in orbit within the next twenty years, and if they can be maintained at reasonable cost the price charged for using them ought to come down. Much depends upon political developments, but at any rate the communications satellite has come to stay.

When we turn to pure science, the scope is just as great. Satellites have already told us more than we have ever known before about the figure of the Earth, for instance; and our knowledge of the upper atmosphere has increased out of all recognition. This may seem to be a purely academic matter, but it is nothing of the kind, because we have long since reached the stage in which every branch of science is linked with every other branch; there are no longer any watertight compartments. Of special interest to astronomers are the OAO vehicles, or Orbiting Astronomical Observatories. Once above the atmosphere, their instruments can be used to study all the radiations coming from space, and not only those which can filter down through the air toward the ground. For instance, there are stars which send out vast amounts of energy in the form of X-rays, and these can be studied only from "up above". There have already been many satellites designed to make special investigations of the so-called cosmic rays (actually, high-speed particles) which come from deep space, and whose origin is still very much of a mystery. Before 1990 there should have been really striking advances in the astronomy of regions beyond the Solar System. We are also learning more about our own local areas, since satellites can, and do, collect samples of interplanetary débris.*

These are only some of the many uses to which satellites will be put between now and 1990. I have not even mentioned vehicles such as navigational satellites; and there is, too, the all-important matter of what is usually termed "spin-off". Miniaturized equipment has to be developed for satellite instrumentation, when it is vital to cut down the weight as far as possible, and the results have been widely adapted, which is something else which the anti-space brigade conveniently overlooks.

To sum up: by 1990 the number of satellites should have been increased very considerably. Many of the vehicles will be in permanent orbits, and will be regularly serviced. Others will be sent

---

*This, too, is something which has been done many times already. One early sample was collected on behalf of a British research team, and was duly delivered for analysis after having been collected from interplanetary space. Unfortunately, British Rail managed to lose it somewhere between London and Crewe.

up for specific tasks, and will be brought down again when their programmes have been completed. We may hope that the really useful satellites will be under international control, if only because all humanity can benefit from these man-made moons.

3

SHUTTLES AND SPACE~STATIONS: 1976 TO 1990

In 1965 Cosmonaut Alexei Leonov, of the Soviet Union, became the first man to go outside an orbital vehicle and "walk" in space. Less than four years later Colonel Thomas Stafford, of the United States, commanded the second probe to make a trip round the Moon. In July, 1975 the two men met "in space", during the first combined mission involving a Russian Soyuz and an American Apollo vehicle orbiting the Earth.

In the first decade of space research there was very little direct contact between the programmes of the two super-powers, while the efforts of other nations were so puny in comparison that to all intents and purposes we can ignore them. Much was heard of the so-called space race, and according to the popular Press there was frantic competition to see who would be first on the Moon. Personally I never had any faith in the space-race idea, as I said many times; the Russian programme was basically different from the American, and relied much more heavily upon automatic probes, at least with regard to the Moon. However, that there was some rivalry there can be no doubt at all, even if it affected the politicians more than the scientists, and a full or even partial interchange of ideas was noticeably lacking. One result of this was that when the idea of co-operation was first mooted, it was only too clear that the two types of space-craft were not suitable for marriage. Put into the vernacular, Russian nuts did not fit American bolts, or vice versa. Even the atmospheres used by the teams were different. The Apollo travellers breathed pure oxygen at a pressure of 5 lb. per square inch; inside a Soyuz the mixture was of oxygen and nitrogen at sea-level pressure of 14.7 lb. per square inch. Any attempt to pass straight through a docking tunnel from one environment into the other would have been disastrous.

There were tragedies, too, during the early period; three American astronauts were killed when their capsule caught fire during a ground test, one Russian (Colonel Komarov) died when his parachute breaking system failed during a return from orbit, and, of course, in 1971 the crew of Soyuz 11 were found to be dead when they landed after spending more than three weeks aboard Salyut, Russia's first space-station. In retrospect, it seems at least possible that all these fatalities could have been avoided if the Soviet and American space teams had been able to pool their information without political interference.

It was therefore all the more remarkable (and the more encouraging) that despite the continued lack of trust between Washington and Moscow, plans could be made for a space link-up in 1975 or 1976. Once the decision had been taken, events moved with commendable speed. There was much to be done; for instance the selected Americans had to learn Russian, while the cosmonauts set out to make themselves fluent in English. It was agreed that the Soviet vehicle should be a modified Soyuz, while the American craft would consist of the command module and the service module of an Apollo – the former to house the crew and the latter to provide propulsion and electric power. Also needed was a docking module, American-built, to serve as an airlock and transfer corridor between Soyuz and Apollo; it was to be around ten feet long and five feet in diameter, and would be cylindrical in shape.

Tremendous care was essential here, because of the atmosphere problem. According to the original plans, the Soyuz pressure would be reduced to 10 lb. per square inch when the docking module had been fixed, so that the Russian cosmonauts could transfer from one craft to the other without taking time in the airlock to breathe pure oxygen and force nitrogen into their blood.

When the flight plans were announced, they were straight-forward enough – in theory! Soyuz would be launched first, into an orbit just under 170 miles above ground level. Apollo would follow a few hours later, taken up by a Saturn 1B rocket, and put into an initial orbit at 130 miles. Apollo would then be separated from its launcher, and the docking module extracted; the Americans would then manoeuvre Apollo up to Soyuz, and dock. While the craft were linked, the crews could pass from one section into the other. After the end of the joint experiments, two days later, the vehicles would separate before landing independently in the usual way.

The agreement for the programme was signed by President Nixon and Mr. Kosygin on 24 May 1972 – a red-letter day in the history of space research. In July 1975 the experiment was duly carried out with complete success – apart from the fact that the Apollo astronauts were affected by poisonous gas during their final descent, luckily with no permanent effects. The names of the space-men should be recorded here. For America, Thomas Stafford, Vance Brand and Donald Slayton: for the USSR, Alexei Leonov and Valeri Kubasov.

Despite its supreme importance, the link-up was not a prolonged affair, and it was only one more step on the road toward a space-station far more elaborate than either Salyut or Skylab. Bear in mind, too, that neither of these early stations had more than a limited useful life. Skylab accommodated three crews in succession, each of three astronauts, and after the departure of the last team (Carr, Gibson and Pogue) in the spring of 1974 it was abandoned; it remains in orbit, but unless plans are made to preserve it or to salvage it the station will eventually fall back into the atmosphere and will be destroyed. A full-scale space-station, on the other hand, will be permanent.

Expense is always a problem, and up to now every flight above the atmosphere has been enormously costly. It takes a very large launcher to send out a relatively small payload. This is shown by the example of Apollo. The original launching-pad height of the whole vehicle was over 360 feet, which is equal to the height of St. Paul's Cathedral; only a 22-ft. cylinder survived to splash down in the ocean, bringing the astronauts home. Moreover, vehicles of the Apollo or Soyuz type can be used only once. The situation is rather like that of a businessman who lives in Brighton and works in London, but who has to buy a new car for each journey if he wants to commute by road.

If frequent space journeys are to become financially possible, the only solution is therefore to develop a vehicle which can be used over and over again. This is where the Shuttle comes in. It is already on the drawing-boards; it should be tested in 1978, and unless there are any unforeseen delays it will be fully operational by the early 1980s. It will cut the cost of a launch by nine-tenths, and it will also be able to carry passengers who are not trained astronauts.

Obviously it will be a compound vehicle, made up of two stages: a booster and an orbiter. It will take off like a rocket, fly in orbit like a space-ship, and land like an aeroplane. According to the first design, the orbiter stage looks not unlike an ordinary delta-wing aircraft. At launch the two parts will be joined together, with the orbiter in a sort of piggy-back position, and separation will take place once the Shuttle is above the top of the resisting layers of atmosphere.

The Russians have not announced any similar design studies, but there is no reason to doubt that something of the same kind is very much in their minds. Whether or not there will be a combined USA–USSR Shuttle programme remains to be seen – probably the results of the space-link experiment will be important here – but at all events it is not likely that any space-station will be a practicable proposition without a recoverable ferry for journeys to and from the ground. Also, assuming that I am right in fore-casting that there will be many highly complicated unmanned satellites circling the world by 1990, and that these will need regular maintenance, Shuttles will be absolutely essential for this reason alone.

What, then, of the space-station itself? Opinions here have swung sharply to and fro. Konstantin Tsiolkovskii was talking about orbital stations as long ago as 1903 (even before, if we take his early novels into account). Then, in the 1920s and 1930s, elaborate designs were put forward by various pioneers. Wernher von Braun, for instance, favoured a wheel-shaped structure, with the crew living in the rim; slow rotation of the wheel would create "artificial gravity" in the form of what is usually termed centri-fugal force. The von Braun wheel design was still to the fore in the years immediately after the war, but then came a period when space-stations fell into disfavour; it was said, for instance, that they could not safely move within the radiation zones round the Earth which had been discovered by the American team headed by James van Allen. By now the pendulum has swung again, but the wheel form has been abandoned, and the modern concept of a space-station is very different. It will not be launched in one piece; this would be out of the question, and indeed there was never any suggestion that an orbital vehicle could be built on Earth and then fired bodily into space. It will have to be assembled beyond the atmosphere.

Here we come back to the Shuttle. Quite probably a space-station of the 1990 period will be made up of a cylindrical "nucleus", made up from one of the original modules, with fixed "off-shoots" as shown in the sketch. If it moves in a roughly circular path at a height of 300 miles or so, it will be to all intents and purposes

40

beyond the range of atmospheric drag, and in any case it will have its own motors which can enable it to be manœuvred around to some extent at least. The power source will be nuclear, though solar panels will be used as well; if the sunlight is there, why waste it? Neither is there any reason why the station should not rotate, though some parts of it will have to stay under conditions of zero gravity so that special experiments can be carried out.

I have already said something about the practical uses of unmanned satellites. All I need really add is that anything which an automatic satellite can do, a manned space-station can do better; the human brain is still the best of all pieces of scientific equipment, and, unlike a machine, it can reason and make on-the-spot decisions.

How many space-stations will be in orbit by 1990? I suspect that there will be at least two, and probably three, though this depends very much upon the success of the Shuttle and upon the overall international situation. As the stations move across the sky they will become as familiar as aircraft are today (and let me add that unlike aircraft, they will be silent and will leave no pollution in the air). The profession of "astronaut" will become less exclusive than it is at present, when the first qualification is that a candidate must be either a Russian or an American. Finally, there is no reason to suppose that space-station crews will be exclusively masculine.

One point about which we are not yet clear is the length of time that a crew member can stay aboard without suffering lasting ill-effects. We know that three months under zero gravity is tolerable; Skylab has shown that. On the other hand, it would probably be unwise for anyone to stay up for as long as a year, even if the station's spin creates a certain amount of apparent weight. The safety period must be balanced against the cost of too-frequent crew changes. I do not want to give the impression that the Shuttle will be cheap. Nothing can make it so, and each flight will cost a great many thousands of pounds even when the vehicles have been perfected.

There must be psychological problems, too. With a space-station containing several dozens of people these problems will be lessened, because everyone will belong to what is virtually a self-contained colony, but there must be ample provision for recreation, together with as much privacy as is possible.

As time goes by, the space-stations will become more and more firmly established, and by 2025, the end of the period I have set out to discuss here, people in general will be hard pressed to

41

remember a time when they did not exist. Moreover, they will have a major rôle to play in preparations for sending men to Mars; they will, in fact, be regarded as marking the Earth's true frontier.

It is more than likely that some of the younger readers of this book will themselves become members of space-station crews. If so, I hope they will cast their minds back to what I have said – and, if I am still alive, let me know how right or how wrong my forecasts proved to be!

4
MAN ON THE MOON: 1969 TO 1995

The Moon, at its distance from us of roughly a quarter of a million miles, seemed to be a very remote world when I began taking a serious interest in it more than forty years ago. Actually, an aircraft which flies ten times round the world covers a distance greater than that between the Moon and ourselves; and when I looked recently at the mileometer of my ancient but invaluable Ford Prefect I found that it registered a grand total of over 600,000 miles – equivalent to more than a trip to the Moon and back. But close though the Moon may be, reaching it is still a very hazardous business, and to my mind it is remarkable that up to the present moment there have been no tragedies in any of the flights there. Nine crews have now been either round or to the Moon.

This brings me at once to an important point. The Apollo plans included no provision for rescue. An astronaut stranded on the Moon would have to stay there, probably in full communication with the rest of mankind but hopelessly out of reach. Had Apollo continued for another ten or twenty missions, then sooner or later something would have gone disastrously wrong. This is why I personally hold the view that it was right to end the programme with No. 17. Before man goes back to the Moon, he must have a vehicle which is much more adaptable.

However, I am running somewhat ahead of my theme, and to put matters in perspective it seems wise to backtrack a few years – in fact to 1958, which seems æons ago now.

It was then that the Americans made their first attempts to send vehicles to the neighbourhood of the Moon, but this was the period when they were having all sorts of difficulties, and there was

LUNA 13

One of the early Russian soft-landing lunar probes: Luna 13. The shadow of the vehicle on the Moon's surface is spectacular. It was these soft-landing probes which showed that the Moon has a surface firm enough to support the weight of even a massive space-craft.

a succession of abject failures. It was left to the Russians to make the first lunar flight. On 4 January 1959, Lunik 1 passed within five thousand miles of the Moon, and sent back some valuable information, notably a confirmation that the lunar magnetic field was either very weak or else absent altogether. On 13 September Lunik 2 made a crash-landing on the surface, an event which deserves to be remembered inasmuch as it was the first direct contact between our world and another. Less than a month later Lunik 3 began an epoch-making flight. It went right round the Moon, and sent back the first pictures of the mysterious "other side" which is always turned away from the Earth, and about which nothing definite was then known.

(*En passant:* "Lunik" was the common term at that time, though after the fourth vehicle in the series it was tacitly replaced by "Luna". Since my knowledge of the Russian language is confined to "Niet" and "Spasse-bo", I cannot pass judgement either way, and frankly I do not regard the matter as important!)

I have vivid memories of Lunik 3, because I was actually in a BBC television studio, presenting a live "Sky at Night" programme,

44

when the pictures came through in my screen, and I had to do some very rapid interpretation. By modern standards, of course, the pictures were very blurred indeed, and some of the features named by the Russians have proved to be non-existent; the classic case is that of the long and imposing Soviet Mountains, which do not exist at all. Yet there were some objects which could be positively identified, notably the immense, dark-floored crater which has been very properly named in honour of Tsiolkovskii. The most important factor was that, as expected, the Moon's far side proved to be just as wild, just as crater-scarred and just as lifeless as the side we have always known. Also as expected, there were fewer of the dark "mare" or sea-surfaces.

After Lunik 3 there was a lull, punctuated only by sporadic efforts by both the Russians and the Americans to do something better, such as obtaining really good photographs from close range. The next successes were from Cape Canaveral, with the last three Rangers, which also crash-landed on the Moon but which

SURVEYOR 3 FROM ORBITER 4
A classic photograph, taken before the first men reached the Moon. The automatic probe Orbiter 4 recorded a space-craft, Surveyor 3, which had previously made a soft landing on the Moon. Surveyor is shown in the box at the middle of the photograph.

sent back excellent pictures during the last few minutes of their flights. In 1966 there followed Luna 9, again a Russian craft, which made a soft landing and continued to transmit information after its arrival; this, incidentally, gave the coup de grâce to a strange but much-publicized theory, backed by the full authority of some eminent professional astronomers, to the effect that the lunar maria were oceans of soft dust into which any space-craft rash enough to land would inevitably sink. (I have never understood why this weird idea was given any credence; elementary observation should have been enough to show that it could not be true.) Then, between 1966 and 1968, there came further soft-landers, together with the five immensely triumphant U.S. Orbiters which went round and round the Moon, providing splendidly detailed photographs of almost the entire surface. By the end of the Orbiter series, lunar mapping – began by men such as Harriott and Galileo during the reign of King James I – was to all intents and purposes complete.

The stage was set for events which would be even more spectacular, but by this time there had been a sharp divergence in the paths of the Russian and the American lunar programmes. Had there ever been the prospect of a race to the Moon, which I personally doubt, the Russians had certainly abandoned it in favour of exploration by unmanned probes. Meanwhile, Apollo was under way, and Christmas 1968 saw three astronauts in Apollo 8 circling the Moon in an orbit which carried them down to a mere 69 miles above the surface. In May 1969 Colonel Stafford commanded Apollo 10 in a test of the lunar module vehicle in the immediate vicinity of the Moon; and on 21 July Neil Armstrong and Edwin Aldrin stepped out on to the Mare Tranquillitatis. (Let us not forget Michael Collins, the only member of the Apollo 11 team who did not go the full way. Without him, the mission could never have been carried through.)

I have never believed that anything can again recapture the excitement of that night; I suppose the first landing on Mars may do so, but I am not sure. There could be no doubt whatsoever of the surge of enthusiasm all over the world, which makes the later public apathy even more puzzling. As Armstrong and Aldrin "found their feet", experiencing the strangeness of one-sixth g, the scene was reminiscent of science fiction; it was all the more striking because we on Earth could watch it and listen to what the astronauts were saying. Yet I remember looking at the grounded Eagle module, and realizing that the single ascent engine represented the explorers' only link with their home planet.

Various bogeys were laid at once. Moving around on the Moon

46

was by no means difficult; communications were excellent; the space-suits functioned perfectly. Neither were there any crises on the return trip. Purely as a precaution, the astronauts were put into strict quarantine on their arrival back on Earth; and this was, in principle, wise. The chances of their bringing back anything harmful were many millions to one against, but when embarking upon a completely new type of activity one cannot be too careful – and, remember, quarantining will be absolutely vital when we come to bringing back samples from Mars, which, hostile though it may be, has an atmosphere of sorts and may not be so sterile as the Moon.

Since I am not writing an account of the past, I will say very little about the remaining Apollo missions, which ended with No. 17 in December 1972. There are only two points which seem really relevant here. Apollo 13, sent up on 11 April 1970, nearly ended in disaster; on the outward journey there was an explosion which put the all-important service module (the space-craft's power-house) permanently out of action, and it was only by a combination of courage, skill and luck that the astronauts were brought home safely. Had the explosion happened on the return journey, nothing could have been done; the lunar module – whose engines saved the day – would have been left behind, and Apollo 13 would have been a helpless hulk. During the period when things were looking really serious, all national feelings seemed to vanish, and messages of goodwill poured in to Mission Control from, I think, practically every country in the world. It was a reminder that wherever we may live, we are all citizens of the Earth; and I do not believe that any other single event in world history has brought this out quite so clearly.

The other relevant point is that with Apollo 17, a professional scientist was included for the first time. Dr. Harrison ("Jack") Schmitt was a geologist who had been trained as an astronaut, not an astronaut who had been given some training as a geologist. It was a foretaste of things to come.

Yet by this time the "anti-space" faction was in full cry once more, and the enthusiasm of 1969 was ebbing away. In the mass media, Apollo 17 received less coverage than football matches and the usual dreary political squabbles. There seems to have been a general opinion that this was one reason why the lunar programme was cut short, instead of continuing through to Apollo 21. The true facts, however, were very different.

The plain truth was that Apollo had done about as much as it ever could. Recording instruments had been set up in six different sites, and of these all but the first were still transmitting information at the beginning of 1975. Further Apollos could only repeat the same procedures in other parts of the Moon; and while this would have been scientifically useful, it did not seem worth risking human lives to achieve it, particularly as it had been shown that limited samples could be obtained by using automatic vehicles. Here the Russians led the way. They sent probes to the Moon and brought them back; they dispatched two Lunokhods or Moon-crawlers, which looked like demented taxicabs but which proved to be superbly efficient. If the Moon could be explored in such a manner, what was the point of sending men there?

The argument was weak, because in fact both programmes were absolutely necessary for any future development of full-scale lunar bases. It was not a question of "either . . . or". Let us, then, take stock of the situation as it is at the present time (1975), and see what is likely to happen next.

During the coming decade – perhaps for even longer – I am convinced that there will be no official plans to send more manned expeditions to the Moon. The Americans have other things to do; there is the all-important Shuttle programme, and there are the Mariner and Pioneer probes to the planets. These ventures cost money, and although the expenditure is very small compared with the yearly outlay on, say, tobacco, it is something that has to be watched, certainly from the White House and quite probably from the Kremlin as well. There is little point in dispatching further astronauts until they can carry out completely new programmes – and, above all, until the risks can be minimized as far as possible.

Essentially, this means using a vehicle which is far more reliable and far more efficient than Apollo. Chemical propellants have marked limitations, and we must await the development of nuclear rockets, which is why I have dated the present chapter to 1995 instead of 1990. Once the space agencies have vehicles which can take men to the Moon and bring them home again without the present-day clumsy and wasteful system of regarding most of the rocket as expendable, the whole situation will alter dramatically. A nuclear rocket of the type which should become operational within the next twenty years ought to be able to manage this.

But as yet, the reliable nuclear rocket lies well ahead; and in the meantime, unmanned exploration of the Moon will go on. Here we may, I think, expect the Russians to take the initiative. Following the successes of their two Lunokhods and their recoverable Lunas, I do not anticipate that they will leave the Moon alone for very long. They are anxious to unravel as many of its secrets as they can, and, unlike NASA, their space agency does not have to answer to the general public for every financial outlay.

Between 1975 and 1900 the Russians may launch a whole series of soft-landing vehicles, so that they will establish a network of recording stations in different parts of the Moon – perhaps even on the far side, collecting their information by using orbital probes as relays. As yet they have not attempted to set up stations of this kind, but probably it is something which is already on their list. Crawling Lunokhods, too, will be dispatched, and there will be more Lunas which can collect Moon-rock and bring it home, although this latter programme may have rather lower priority.

If the network is to be operational by, say, 1983 we may expect many launchings during the previous year or two. I also suggest that the Soviet planners may send several payloads to one specially selected area (perhaps the Mare Imbrium?) so that when their pioneer cosmonauts arrive, at a later stage, they will find plenty of miscellaneous material to hand. Remember that on the Moon, with its lack of atmosphere, there is no "weather", and there is no reason why materials sent from Earth should deteriorate, though the effects of various kinds of radiations have to be taken into account.

Before turning back to Cape Canaveral, what about other nations? I hardly expect much in the way of lunar exploration from Europe, excluding the U.S.S.R., but things may be different when we consider the Far East. The Japanese excel at "cut-price" miniature probes, and they are quite likely to join in by sending small vehicles moonward. The Chinese interest (if any) is much less easy to evaluate, but it may be that they are more interested in Earth satellites. All the same, Chairman Mao, or whoever succeeds him, might want to send a space-craft to the Moon merely to show that China is not being left behind.

The only definite probability so far as America is concerned, for the period until now to the early 1980s, is a polar orbiter. Though most of the Moon's surface has been effectively photographed, there is still a small area, near the south pole, which has not been fully covered to the same degree of accuracy, and this is one reason why scientists are interested in a vehicle orbiting the Moon over both polar regions. There are other reasons, too; it is important to find out what conditions near the Moon will be like at the time of the next sunspot maximum, which is due around 1980. If funds are forthcoming, the polar orbiter is scheduled for launching some time during 1979.

Incidentally, it is true to say that many of the thousands of photographs sent back by the five Orbiters of 1966–1968 still remain to be fully studied. Some of them have not been analyzed even casually, and remain in their boxes. Of course, many more – including those of surpassing interest – have been released; but a full study of all the pictures will take years, so that the analysts already have more work to do than they can conveniently handle. It is rather ironical that despite all the data collected, we still do not know nearly as much about the Moon as might be thought. The argument about the origin of the main craters and walled formations – vulcanism, impact, or something different from either – still goes on, and we remain rather in the dark about the make-up of the lunar globe below the outermost, fragmented layer or "regolith".

Each Apollo mission deposited an ALSEP, or Apollo Lunar Surface Experimental Package. This involved the setting-up of equipment by astronauts during their stay on the Moon, and I repeat that apart from the preliminary ALSEP of Apollo 11 all the stations are still working well. Whether the Americans will establish further transmitting stations, using automatic vehicles, remains to be seen; but in the immediate future it does not seem likely.

So far as the Moon is concerned, then, I predict that the only major activity for the next fifteen years will be the continuation of data-collecting by Russian Lunokhods, the possible establishment of a network of surface transmitters by the Soviet teams, and at least one polar orbiter from America. The main U.S. emphasis will be on Shuttles, space-dockings and orbital bases. But if nuclear rockets are developed as quickly as we are entitled to hope official attention will start to swing back toward the Moon by 1990, and we may expect a significant change in attitude.

There can be no doubt about the value of a Lunar Base, but before practical construction can begin there must be further manned reconnaissances. Bear in mind that by 1990 none of those who have been to the Moon will be young enough to make a return trip. In fact, by 1975 there were only two of the "Moon-men" still on NASA's flight status list: Admiral Alan Shepard and General Thomas Stafford. Shepard was forty-seven years old when he went to the Moon, which means that by the time lunar flights are recommended he will be verging upon seventy; and this is really over the age limit, even for a man of his remarkable ability and stamina. Neither will Stafford then be still young enough to take an active part. There must be a new generation of lunar astronauts; much of what was learned from Apollo will have to be re-discovered.

All in all, I expect that in the years following 1989 or 1990 there will be new expeditions, using new types of rocket vehicles which will make Apollo seem very antiquated. It is very much to be hoped that the trips will be international, and that all the results will be pooled. Some of the missions will last for much longer than the brief there-and-back hops made at the time of Apollo, and preparations will include the transport of materials which will be needed for the Base itself. This phase should end before 1995, and the next great step forward will be imminent.

It seems a long time ahead, but there is a great deal of sound common-sense in the old cliché about learning how to walk before trying to run, and the Moon, at least, is content to wait.

5
PROBES TO THE INNER PLANETS: 1975 TO 1990

The first planetary probe in history was launched by the Russians on 12 February 1961. It was not a success. It was meant to investigate Venus, and may have by-passed that peculiar planet within 65,000 miles or so; but since all contact with the probe was lost at an early stage, and was never regained, nobody will ever know just what happened to it. Less than two years later America's Mariner 2 approached Venus within 22,000 miles, and sent back the first reliable results from the immediate neighbourhood of an alien world. Since then we have also had missions to Mars, Jupiter and Mercury, and there is no reason to doubt that before the end of the century our automatic probes will have penetrated to the very boundary of the Sun's main system.

Moreover, Nature has provided us with an exceptional opportunity inasmuch as the outer planets are conveniently arranged for what I have unromantically called "interplanetary billiards". Whether the chance will be taken is another matter, but I will defer a discussion of it until I come to talk about the probably ill-fated Grand Tour concept.

Obviously, contacting the planets is by no means so straightforward as sending vehicles to the Moon, because the planets do not stay pleasingly close to us. They move round the Sun, not round the Earth, and in any case even the nearest of them (Venus) is always at least a hundred times as remote as the Moon. Almost everyone today can list the planets in their order of distance from the Sun – just in case of any confusion, I have given an over-simplified scale diagram – but it may, I think, be better if I give them here in the order of their minimum distances from the Earth, reduced to round numbers, and remembering that the closest we can ever approach to the Sun is $91\frac{1}{2}$ million miles.

So here is the roll-call of the Solar System:

Planet	Distance from Earth, in millions of miles		First successful probe contact
	Minimum	*Maximum*	
Venus	24	162	1962: Mariner 2
Mars	34	250	1965: Mariner 4
Mercury	48	137	1974: Mariner 10
Jupiter	366	600	1973: Pioneer 10
Saturn	743	1030	? 1979: Pioneer 11
Uranus	1600	1960	? 1985, a Mariner
Pluto	2660	4950	? About 1988
Neptune	2675	2910	? About 1988

Pluto, let us note, has an unusual orbit which is both exceptionally eccentric and exceptionally tilted. At the moment it is drawing in toward perihelion, which it will reach in 1989, so that for some years to either side of that date it will temporarily forfeit its popular title of "the outermost planet". There are all sorts of problems connected with Pluto, which may really be nothing more than an ex-satellite of Neptune which somehow achieved a cosmical U.D.I. and is masquerading as a planet in its own right. But before going into further detail, it is important to look at the general scheme of planetary exploration, because the journeys take much longer than might be thought.

What cannot be done, with our present-day vehicles, is to wait until a planet is at its closest to us and then fire straight across the gap. This would mean using power almost all the way, and no space-craft could carry nearly enough propellant. The procedure is to take the probe out beyond the Earth's atmosphere and then either accelerate it or decelerate it, depending upon the selected target. To decelerate the probe, relative to the Earth, means that it will start to swing inward toward the Sun, and can be made to rendez-vous with Venus or Mercury (or both, one after the other, as with Mariner 10). Accelerating the probe will swing it outward toward the path of Mars or one of the more remote planets. If a landing is planned, or the vehicle is scheduled to go into a closed path around its target, extra power will have to be applied at the critical moment – and at least one mid-course correction must usually be made in any case. Everything has to be "just right", since an apparently

53

TRANSFER ORBITS

Because of propellant restrictions, it is not (as yet) possible to send a probe from the Earth to another planet by a direct route. The probe has to be either accelerated or decelerated relative to the Earth, so that it will enter a 'transfer orbit' and rendezvous with its target

planet, moving for most of the time in unpowered free fall. In this
diagram a probe is shown moving from the Earth inward toward the
Sun, making rendezvous first with Venus and then with Mercury.
The first probe of this kind was Mariner 10, which by-passed Venus
in February 1974 and Mercury in the following month.

insignificant error in either velocity or direction means that the probe will miss its target by a great many millions of miles.

It is probably true to say that the early plans for sending vehicles to the planets were purely scientific, with no military undertones, and this is still the case today; I doubt whether even the most devious politician can see any tactical advantage in finding out whether Mercury has volcanoes on its surface or whether Jupiter's Great Red Spot is the top of a whirling storm. This is excellent in its way, but it means that the space-planners have to fight even harder for the necessary funds, and already there have been some crippling economies, of which the most depressing is the cancellation of the Grand Tour in its original form. However, things might be worse, and at least nobody can deny that the space-craft launched between 1962 and 1975 provided an immense amount of fascinating new information.

Before the probe era began, our knowledge of the planets was really somewhat fragmentary – more fragmentary, indeed, than most people believed at the time. Around 1960 I remember giving a lecture to a scientific society in which I described conditions on the surface of Mars to my entire satisfaction. Mars, I said, was a world with a moderately dense atmosphere made up chiefly of nitrogen; it had white polar caps composed of a thin layer of ice or frost; the landscape tended to be uniform, with no major mountain ranges anywhere; and the famous dark areas, visible in a small telescope under favourable conditions, were depressed basins, probably old sea-beds and certainly filled with low-type vegetation. Alas, every one of these statements has been shown to be wrong, and Mars is emphatically not the sort of planet which we had expected it to be. Our views about it have changed more in a decade than they did in the previous century.

A long-term project in the minds of many people was the establishment of colonies on at least some of the planets, beginning presumably with either Mars or Venus. (When the Space Age began, it was often thought that Venus might be the more promising of the two.) One suggestion, however, could be ruled out of court at once. There is no hope of solving the Earth's over-population problem by extra-terrestrial colonization. I am fully prepared to believe that within the next thousand years there may be a million people living on the Moon and even more on Mars; but what is this when we consider the vast population of the Earth itself, quite apart from the steady and sinister increase in the birthrate? As I have commented elsewhere, it is rather like trying to solve London's traffic problem by banning all cars registered in

Bognor Regis. The problem is relevant here, because quite apart from what happens on Earth it will be absolutely necessary to control the population numbers on the Moon or Mars; but this is something to be discussed when we turn to manned bases, and for the moment let us confine our attention to automatic vehicles.

As with the satellite programmes, there has been a sharp difference between the Russians and the Americans in their mode of attack. The Soviet approach has been dominated by their attempts to soft-land probes capable of sending back information after arrival, while the Americans were more concerned with fly-by missions all through the 1960s and into the early 1970s. Moreover, the Russians have had triumphant success with Venus and practically none with Mars, while the Americans have yet to soft-land a vehicle anywhere except on the Moon, though when it comes to long-range control and photography the United States technicians are vastly better than those of the U.S.S.R. Once again we come back to the folly of not pooling knowledge.

Before 1962 Venus was usually nicknamed "the planet of mystery", and opinions about it differed widely. It is about the same size and mass as the Earth, with an escape velocity which is only slightly lower; but it is hidden beneath a dense, cloud-laden atmosphere, and in fact nobody has seen the true surface of Venus even yet. There were two main theories. The first dismissed Venus as a raging dust-desert, with an extremely high surface temperature and a total lack of water; the second regarded the surface as mainly oceanic, with almost no dry land at all. The atmosphere was known to be made up principally of carbon dioxide, at least in its upper layers, and the rotation period was thought to be about a month – remembering that Venus' "year" is equal to 224.7 of our days.

Of these two pictures, the second was much the more attractive, and it opened up the possibility of primitive life there. In its youthful period the Earth had much more carbon dioxide in its atmosphere than is the case now, and there were oceans in plenty; life began in the seas, and gradually developed. Only when land plants gained a foothold did the atmosphere change, since these plants removed carbon dioxide from the air and replaced it by free oxygen, according to the principle of what is termed photosynthesis. It seemed at least possible that Venus might be a world in an early stage of development – and if so, then life could presumably evolve there just as it has done here. A rather bizarre consequence would be that if the carbon dioxide entered the seas, Venus would have oceans of soda-water. The dust-desert idea was

57

Uranus

Neptune

Pluto

DIAGRAM OF THE SOLAR SYSTEM
The Solar System contains nine known planets, at different distances
from the Sun. This diagram (which is not to scale) shows the orbits,
from that of Mercury out to that of the most remote planet, Pluto.

much less inviting, and most people hoped that it would prove to be wrong.

This was not to be. Mariner 2 reported a scorchingly high surface temperature, together with a very slow rotation of around 243 days, which is longer than Venus' "year". Even more surprisingly, there was no detectable magnetic field. Later probes gave the same story. In 1967 Russia's Venera 4 made a controlled parachute descent through the planet's atmosphere, though apparently the great heat and pressure put it out of action before touch-down; since then Veneras 7 and 8 have maintained radio contact for some time after arrival, and we must now relegate those intriguing seas of soda-water to the realm of fiction. Under a temperature of 900 degrees Fahrenheit no liquid water could exist – despite the atmospheric pressure, which is of the order of a hundred times that of our own air at sea-level.

Meantime, radar techniques had been introduced, not from space but from ground level. American research teams "bounced" radar pulses off Venus, and established that large, shallow craters existed there, at least in the regions which were examined. The stage was set for Mariner 10, which was the first probe to take in two target planets instead of only one.

Mariner 10 made use of the interplanetary billiards technique. It was launched in the usual way, and allowed to swing in toward Venus, which it by-passed on 5 February 1974 at a distance of 3000 miles. For the first time, close-range pictures were obtained, confirming that although Venus itself is a slow spinner the top of the cloud-layer rotates relatively quickly in a period of four days, indicating a most remarkable atmospheric structure. As Mariner approached Venus, the planet's gravitational pull altered the path of the space-craft, and slowed it down relative to the Sun. This meant that it continued to swoop inward, and on 29 March it achieved a rendezvous with Mercury, the innermost planet. Let it again be stressed that everything had to be very accurate. An uncorrected one-mile error at Venus would result in a thousand-mile miss of Mercury.

The Mariner 10 pictures of Venus showed the belts of high cloud very clearly, but of course it was impossible to see through to the surface, and we have to admit that Venus guards its secrets well. There can be no hope of obtaining views from below the clouds, and the chances of sending a manned expedition there are extremely slim, since in its own peculiar way Venus is much more hostile than either the Moon or Mercury. Colonization is unthinkable unless we can do something to alter the environment,

APOLLO 14 ALSEP
The ALSEP (Apollo Lunar Surface Experimental Package) set up in
the Fra Mauro area of the Moon by Astronauts Shepard and
Mitchell, from Apollo 14. The instruments in this ALSEP were still
functioning, and sending back data, in 1975. Shepard and Mitchell
had a 'Moon cart' to transport their equipment; the final three
Apollo missions used LRVs or electrically-driven Lunar Roving
Vehicles.

SOLAR OBSERVATORY ON MERCURY (*overleaf*)
Mercury is so close to the Sun that it will be an ideal site for a solar
observatory. Much of the equipment will be automatic – for the
moment, at least, we cannot suppose that astronauts will be able to
stay on this hostile planet for very long.

JUPITER

Jupiter, as photographed with the 60 in. reflector at Mount Wilson; reproduced by kind permission of the Hale Observatories. When this picture was taken the South Equatorial Belt was relatively thin and obscure (compare with the Pioneer photographs of 1973, when the belt was much more developed).

JUPITER FROM PIONEER 10 (*left*)

Another view of Jupiter taken from Pioneer 10 during the rendezvous in December 1973. The Red Spot, now known to be a meteorological phenomenon, is well shown, as are the complex belts and the bright zones.

JUPITER FROM PIONEER 10 (*below left*)

A close-range view of Jupiter, from Pioneer 10 (December 1973). Note the phase, which is never visible in this form from Earth. The Jovian dark belts and bright zones are well shown. Pioneer made one pass of Jupiter, at a minimum distance of about 80,000 miles; it was fortunate that it did not go much closer, as its instruments would have been put out of action by the intense radiation.

which is a project for the distant future even if it turns out to be feasible at all. On the other hand, Venus is much too fascinating a world to be ignored, and further probes to it are already being planned.

In 1978, if all goes well, a space-craft weighing over 800 pounds

DISK OF VENUS
Venus, photographed from Mariner 10 in February 1974. The atmospheric bands are well shown; note the different appearance near the planet's pole. The upper atmosphere has a rotation period of only 4 days, though the solid body of the planet seems to have a period of 243 days.

will be used to drop four scientific probes on to the surface, after which the main vehicle will itself enter the planet's atmosphere and continue to transmit data until it is destroyed. The aim is not only to see what we can discover about the surface, but also to study conditions immediately round the cloud-layer. There is considerable enthusiasm in NASA circles for further missions, both orbiters and landers, during the 1980s and 1990s, and undoubtedly the Russians will continue in the same vein. One thing to remember is that in order to survive under those crushing, all-concealing cloud layers, any capsule must be extremely tough, and able to resist both heat and pressure. The Russians learned the hard way, but even so there has as yet been no lander able to remain in touch for more than about an hour after arrival.

We cannot be absolutely sure that the whole of Venus is permanently very hot, but all the evidence points that way, and it is significant that there seems to be no measurable difference between the sunlight and the night hemispheres.

All sorts of speculations have been published about the chances of finding any life-forms in the atmosphere of Venus. I can only say that I am profoundly sceptical, because the conditions are intolerable by any standards. Just why Venus is in this state remains unknown, and neither can we account for the strange rotation; the axis is almost upright, and Venus spins round in a direction contrary to that of the Earth. One novel suggestion was that in the remote past Venus was hit by either a comet or an asteroid, which fractured the crust and sparked off extensive vulcanism; this, it is said, explains the thick, carbon-dioxide atmosphere, and also the slow retrograde rotation. The theory is not impossible, but it does seem rather wild. Another recent and rather unwelcome discovery is that one of the minor constituents of the atmosphere is sulphuric acid.

If I may be allowed to digress for a moment, this may be the time to refer briefly to some of the "independent thinkers" who have produced ideas very much of their own. There can be nobody who has not heard of flying saucers, and I well remember meeting the late George Adamski, who maintained that he had become on terms of close friendship with some "Venusians" and had even been taken for a ride in one of their saucers. Equally remarkable is Dr. Immanuel Velikovsky, who is convinced that Venus began its career as a comet and was ejected from Jupiter – after which it bounced around the Solar System in the manner of a cosmical ping-pong ball before having its tail chopped off and settling down to its present sober, steady existence.* Undoubtedly these theories add to the gaiety of nations, if nothing else!

To sum up: Mariner 10 has sent back close-range pictures of Venus' cloud-layer, and several Russian probes have soft-landed on the planet, though all have lost contact rather quickly. The American mission scheduled for 1978 may be more informative, and further orbiters and landers may be expected to follow, but all things considered Venus has proved to be much more hostile than even the pessimists had expected.

Having by-passed Venus, Mariner 10 went on its way toward Mercury, which is a very different kind of world. Meantime, the mission controllers were becoming distinctly nervous. At an early stage in its flight, and quite without warning, the probe had switched from its scheduled power supply to its back-up system, so that it had been using its reserves throughout the journey, and there were serious doubts as to whether the power would last for long enough to complete the programme. However, nothing could be done except to hope for the best.

Mercury is not a great deal larger than the Moon, though it is considerably denser and more massive. From Earth it is very difficult to study, because it always keeps close to the Sun in the sky and shows phases similar to those of Venus – so that when it is at its closest to us it presents its dark or night side, and we cannot see it at all (except on the rare occasions when it passes exactly between the Earth and the Sun, and can be observed in transit as a black spot against the brilliant solar disk). Its "year" is 88 Earth-days, and it was known to spin slowly, completing one turn in $58\frac{1}{2}$ days. Up to 1974, the best map of the surface features remained that of the Greek astronomer E. M. Antoniadi, whose work had been carried out forty years earlier; the map showed dark patches and brighter areas, but very little else. Radar work had indicated the existence of craters, and it was suggested, though without proof, that the landscape might be very like that of the Moon. Life of any kind on Mercury was discounted, both because of the violent extremes of temperature and because of the virtual absence of atmosphere.

To the great relief of everyone concerned, the Mariner 10 cameras lasted for long enough to complete the main programme of studying and mapping Mercury. Closest approach occurred on 29 March 1974, when the probe was a mere 450 miles above the planet. The surface did indeed prove to be crater-scarred, and

---

*I have discussed the Independent Thinkers, including Dr. Velikovsky, much more thoroughly in my book *Can You Speak Venusian?*

MERCURY: CRATER KUIPER
Another photograph of Mercury from Mariner 10, taken from closer
range. The bright, regular crater is known as Kuiper, in honour of
the great astronomer G. P. Kuiper, who died before the mission.
Crater Kuiper is a ray-centre, and intrudes into a larger, less
brilliant crater which also is the centre of a system of bright rays.

indeed the craters were as thickly clustered as those on the most
crowded part of the Moon. The upper layer was also distinctly
lunar in type, and a definite magnetic field was found; there were
also traces of atmosphere, though with a density too low to be of
any practical use.

There was a flurry of excitement during the transmission period
when the ultra-violet experimenters discovered an object which
seemed to have a movement slightly different from that of the planet

MERCURY, FROM MARINER 10
Crescent Mercury; photograph taken in March 1974 as Mariner 10
approached the planet, and for the first time we could study the
lunar-type craters which cover the surface.

itself. Could Mercury have a satellite, after all? It had always been
regarded as a solitary traveller in space, and the detection of an
attendant would have been of the utmost significance. Alas, it soon
transpired that the object was nothing more than a star which
happened to move across the field of view.

The first results were confirmed by two further encounters with
Mercury by Mariner 10 – one in the autumn of 1974 and the other
in the spring of 1975. Of course Mariner 10 is still orbiting the
Sun, but its useful life is over.

For the first time we are now in a position to draw up detailed
maps of Mercury, and this is being done as quickly as possible.*
However, it cannot really be said that our ideas about the surface
conditions have changed much. Mercury is fiercely hot by day,
intensely cold during its long night; the rocks are barren and

*The International Astronomical Union, the controlling body of world astron-
omy, has accepted a suggestion that the first crater to be identified as Mariner
approached Mercury – a bright ray-crater, superimposed on an older structure –
will be named "Kuiper", in honour of Dr. Gerard P. Kuiper, one of the great
pioneers of planetary science, who had died unexpectedly a few months before.

sterile, and the loneliness and desolation there are beyond belief. This being so, it may be asked whether there is any point in making further close-range studies.

The answer is an emphatic "yes", not only because of the scientific interest of Mercury itself, but because of its unique value as an observation site. It is, on average, a mere 36,000,000 miles from the Sun, and during daylight it is exposed to the full fury of the solar radiation. An automatic transmitting station set up there would be highly informative, and there seems no reason to doubt that one will be set up there in the foreseeable future. I suggest that it will be operative by 1985 at the latest, and perhaps earlier – say 1983, though it will have to be very carefully designed and manufactured in order to cope with the amazingly hostile environment. Before then there may well have been a Mercury orbiter, and certainly another fly-by probe to complete the mapping begun so triumphantly by Mariner 10.

The method of bringing a probe down to a soft landing depends upon the atmosphere (or lack of it) round the target planet. Venus provides full scope for parachutes, which have been used for all the landing vehicles so far, though both the rocket braking and the parachute deployment have to be automatic; a command from Earth, even though travelling at 186,000 miles per second, still takes some time to reach Venus. With Mercury, everything depends upon rocket braking, since the atmosphere is much too thin to be noticeable, and the method of landing is basically the same as that for the Moon. Mars, with its thin atmosphere, offers some promise for a parachute, but here too the main emphasis must be on slowing the entry probe down by means of rocket engines fired against the direction of motion.

This leads me on to the Red Planet itself, which has now regained its rightful place as No. 1 prospect for a manned colony beyond the Moon. Though it can never approach us as closely as Venus, we know much more about it, and despite its hostile nature it is much less unlike the Earth than any other body in the Solar System. Neither can we yet rule out the chance of finding some life-form there, though I admit to being decidedly sceptical.

Telescopically, Mars looks very different from Venus. Instead of a featureless crescent or half-disk, the markings on its red globe are (usually) sharp and clear-cut; they can be mapped, and they have been given names. The first nomenclature was, to my mind, rather attractive; we have features such as the Hourglass Sea, De la Rue Ocean, Maraldi Strait, Beer Continent and so on. (Let me add that Wilhelm Beer was one of the two Germans who drew up the first really useful map of the Moon; the other was Johann Mädler.) Then, in 1877, the Italian astronomer Giovanni Virginio Schiaparelli compiled a more detailed chart and revised the list of names, so that the Hourglass Sea became Syrtis Major, and so on. Schiaparelli's system has lasted until the present day, but new revisions and additions are now being made on the basis of the Mariner 9 photographs – of which more anon.

Schiaparelli's work was memorable in another way also. During 1877, when Mars was exceptionally well placed, Schiaparelli described numbers of thin, straight lines crossing the reddish "continents", and it was these which became famous – or, perhaps, notorious – as the Martian canals. Percival Lowell, who set up an observatory at Flagstaff in Arizona mainly to study Mars, believed these canals to be artificial waterways, built by the local inhabitants to provide a vast irrigation system upon a world where water was in very short supply. There emerged a fascinating picture of pipes, pumps, oases and strips of fertilized land. The Martians, said Lowell, must be highly civilized, with a sort of Utopian government.

Lowell died in 1916. Even by then his ideas had been hotly challenged, and it is probably true to say that they were never accepted by more than a minority of astronomers. Other observers, using telescopes every whit as powerful as Lowell's 24-inch refractor, either failed to see the canals at all or else recorded them as vague, streaky features not in the least artificial in aspect. I belong to the latter class, since I have never seen anything in the nature of a hard, sharp line on Mars, even though some of Lowell's canals do have a basis of reality. In any case, this particular problem has now been solved. There are no true canals, and no Martians. Neither are there any huge tracts of vegetation, and the white polar caps are composed mainly of solid carbon dioxide, though there must be a certain amount of ordinary ice as well. If there is any life on Mars, it is of very primitive type.

The first successful Martian probe was Mariner 4, launched in 1964 and which by-passed the planet at just over 6,000 miles in July 1965. Mariners 6 and 7 followed in the summer of 1969, and then, two years later, came Mariner 9, which went into a closed orbit round Mars and sent back thousands of detailed pictures. For the first time we could study the huge craters, the deep valleys and the giant volcanoes. Yet perhaps the most significant result of this work has been the revelation that the Martian atmosphere is painfully thin, with a ground pressure of below 10 millibars. In other words, it corresponds to what we would normally call a vacuum; and instead of being made up of nitrogen, it seems to be composed chiefly of carbon dioxide.

Consequently, the surface of Mars must be more or less un-protected from the various harmful radiations coming from space, and this at once reduces the chance of our finding any advanced life-forms there. Neither can there be any liquid water on the surface, because the atmospheric pressure is too low. (Theoretically, water could persist as a liquid in just one place – the depressed plain of Hellas, which is the lowest-lying area of Mars; but I fear that explorers of the future will have no prospects of finding even a puddle.) Moreover, the climates are somewhat forbidding. According to the latest estimates, the atmospheric temperature near the surface has a range of from +50 degrees Fahrenheit at mid-summer on the equator down to −200 degrees in the middle of a polar winter. Each night is bitterly cold, even at the equator, because the thin atmosphere is most inefficient at blanketing in the Sun's daytime warmth. Any Martian vegetation must be of re-markably tough type.

NODUS GORDII
Nodus Gordii now officially renamed Arsia Mons, one of the loftiest volcanoes on Mars, photographed in 1972 from Mariner 9. It rises to about 15 miles above the general level of the ground, and is therefore much loftier and more massive than any volcano on Earth.

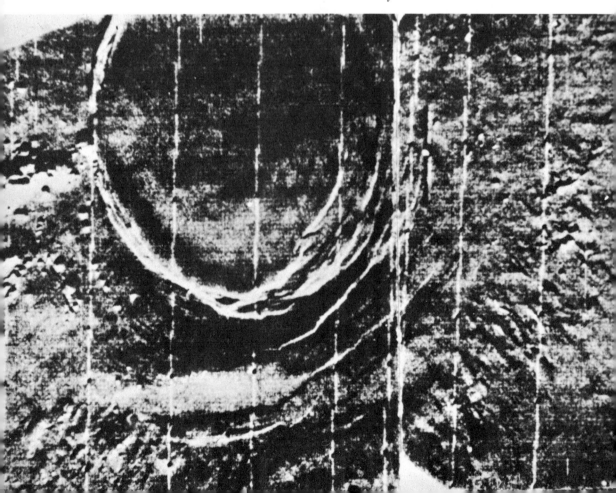

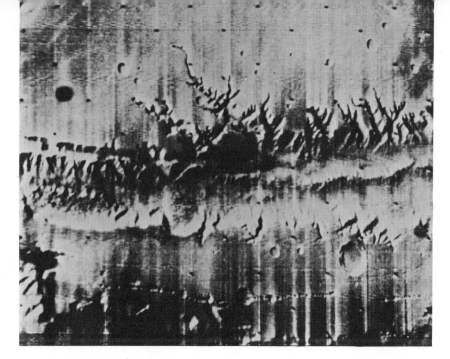

TITHONIUS LACUS/COPRATES

The Coprates Canyon or Valley, near Tithonius Lacus on Mars. This is a huge valley, far dwarfing any structure of the same kind on Earth. It seems to be part of a drainage system associated with the great volcanoes such as Nix Olympica and Nodus Gordii, and gives the impression of having been cut by some liquid (presumably water) though liquid water cannot exist on Mars now, because of the low atmospheric pressure – less than 10 millibars. There are many valleys on Mars, but that of Coprates is the most spectacular of them.

And yet Mars is not inert. Periodically we can see widespread dust-storms there; one of these occurred in 1971, when Mariner arrived in the vicinity of Mars in November, and for some frustrating weeks the space-craft could do no more than photograph the top of an almost featureless dusty layer, though it is true that a few mountain summits poked out. Only when the dust cleared away could the main programme begin. There was another storm in 1973, and there were long periods when the telescope in my own observatory in Selsey showed almost nothing but a blank disk.

The cause of these storms is still a matter for debate. The official explanation is that they are due to dusty material whipped up from the deserts by winds; but I find it rather hard to believe that even a gale in this very tenuous atmosphere could have such widespread and quick results. On the other hand, there are also grave difficulties in the way of supposing that the obscuring material is volcanic ash shot out from still-active vents. Whether Mars is geologically active today remains to be seen. I suspect that it is; but I am quite ready to be proved wrong.

73

Mariner 9 changed all our ideas about Mars. Few people had expected that there would be structures such as Nix Olympica (or Olympia Mons), with its 300-mile base, a 40-mile summit caldera and a total height of about 15 miles above the mean surface level, making the volcano much more massive than anything in our own Hawaii. There is also the highly significant point that many of the winding valleys look so like old riverbeds that in all probability they *are* old riverbeds. Were they really ancient, by our own geological standards, they would be largely eroded away and filled in, whereas they look remarkably fresh. This leads on to a strange paradox. As we have seen, liquid water cannot exist on Mars now; but if it existed in the fairly recent past (that is to say, a few tens of thousands of years ago) the planet must then have made much more atmosphere than it has at the present epoch. There are suggestions that Mars may go through what we may call "fertile" periods,

PHOBOS
Two views of Phobos, as seen from Mariner 9 in 1971. The little satellite is shaped rather like a potato, and is pitted with craters. It has a maximum diameter of less than 20 miles, and is therefore quite unlike our own massive Moon; it and Deimos may well be captured asteroids.

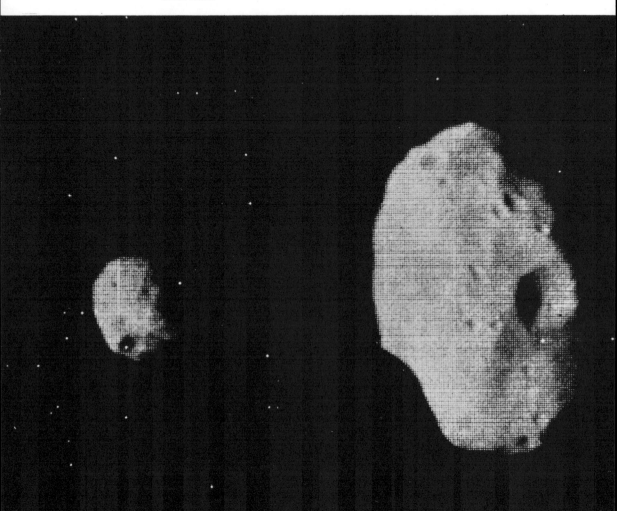

interspersed with periods of sterility. I will have more to say about this later. Meantime, what about the next stages in probe research?

The Russians launched four Mars vehicles in 1973, and evidently intended to soft-land at least two of them in the spring of 1974. Unfortunately they continued to be dogged by bad luck. Two of the probes missed Mars altogether; one went into a closed orbit, but apparently sent back little information of value; and the fourth crash-landed, though contact with it was lost before touchdown and it too must be written off as a failure. The Soviet workers had had similar mishaps with their previous Mars vehicles, which reached the neighbourhood of the planet at about the same time as Mariner 9. One of them dropped a Soviet pennant (which did not seem particularly useful) and the other came down in the area between the regions known as Electris and Eridania, but stopped transmitting a mere twenty seconds after impact. Since I decline to believe that some peevish Martian went up and switched it off, we can only conclude that its radio failed for technical reasons. No doubt it will be found and examined one day, but for the moment nothing more can be said.

Therefore, our main hopes rest with the Americans, and two very elaborate probes are scheduled for launching in 1975, making up the Viking programme. The first should land on 4 July 1976, in the region called Chryse; the planned co-ordinates on Mars are latitude 19.5°N., longitude 34°W. Chryse lies at the north-east end of a tremendous rift discovered on the Mariner 9 pictures; it is low-lying, and several dry rivers seem to converge upon it. There may, then, be "river" deposits there, and since the region lies about three miles below the mean surface level it is a likely place to look for life. The second landing will be at latitude 44°N., longitude 10°W., in the even lower area of Cydonia, at the south edge of the north polar cap. The planned date of arrival is 24 August 1976.

When the landings take place, Mars and the Earth will be well over 200 million miles apart, so that the entire touch-down manœuvre must be automatic; if a last-minute radio command were sent out from Earth, it would take more than a quarter of an hour to arrive. Moreover, the whole procedure is complicated. Each Viking consists of two main parts, an orbiter and a lander, which will be separated only when the vehicle is safely in a closed orbit around Mars. The lander, which will have been sterilized as efficiently as possible in order to prevent carrying any contamination from Earth, will at first be protected by an "aeroshell" or lampshade-like cover; remember that Mars does have an atmosphere, and friction during entry will set up considerable heat

75

which would otherwise damage the instrumentation. At 19,000 feet above the ground, a fifty-foot parachute will unfold above the aeroshell and slow it down; the aeroshell will then be jettisoned, and at about a mile above the surface the parachute will be jettisoned too.

This means that the final descent must be controlled by the rocket motors. The procedure will be not unlike that of an Apollo landing on the Moon, though with the vital difference that there will be nobody on board to take charge if things start to go wrong (and remember that every Apollo touch-down was manually directed by the pilot). When the three footpads touch the surface, the rocket engines will be switched off. The whole period between separation from the orbiter and the final landing will be no more than thirteen minutes.

Obviously there are any number of things which could cause trouble. We can only hope that no hitches will occur; certainly the Americans are well equipped to tackle this kind of operation. It would be infuriating to complete a journey lasting for 11 months, and covering about 450 million miles, only to lose contact at the last moment as the Russians have done.

All sorts of experiments are planned. The orbiting stages will carry out surveys in much the same way as those from Mariner 9; we ought to learn much more about the make-up of the atmosphere, and the photographic results will show up any changes which have taken place since Mariner 9 "went silent" in October 1972. However, the main interest will centre upon the landers, which carry various instruments. One of these is a scoop which will draw in material from the Martian surface to the inside of the probe, where analysis will be carried out and the results reported back; then, for the first time, we should have something really positive to guide us in our search for living organisms. We will also find out what the surface materials are really like, and see whether there is any trace of water, frozen or not. Another device will measure the windspeeds, which may be high even though in that thin atmosphere they will have very little force. Precise temperature-measurements will be made, and seismometers will be deployed in order to detect any possible ground tremors or "Marsquakes".

If all goes well, the landers will have useful lives of about three months after touch-down. The orbiters are designed to operate for over four months, but may last for much longer than their scheduled period of activity, as Mariner 9 did.

The subsequent exploration of Mars is bound to depend very largely upon the Viking results. If there is any hint of life, I believe

76

that research will be financed and pressed forward as quickly as possible, because the discovery of any living organism beyond the Earth would be of the most tremendous significance. Even if not, a further Viking is scheduled for 1979, and there is no reason to doubt that others will follow. The next milestone will then be the recovery of samples from Mars, by means of a probe which goes there, lands, collects the required samples and brings them home, as the Russians have already done with regard to the Moon.

This brings me on to another topic which has already caused considerable anxiety: sterilization. As we have seen, the first astronauts to return from the Moon were strictly quarantined until it had become quite clear that they had brought home nothing harmful. The risks with the airless, sterile Moon were negligible, and the quarantining was more a matter of principle than anything else; but Mars is a different sort of place. We have no proof that it is entirely devoid of life, and we will still be uncertain even if Viking shows no trace of it, since it is always possible that Martian organisms may be confined to a few favourable areas.

Most people will have read H. G. Wells' classic novel *The War of the Worlds*, in which the Earth is invaded by monsters from Mars who proceed to spread alarm and despondency with their heat-rays and other charming devices. In the story, the aliens are eventually killed by Earth bacteria, to which they have no resistance. Now, suppose that a returning space-craft brought back some Martian bacteria which could thrive in our own environment, and proved dangerous to life of terrestrial type? I can hear some readers muttering "Science fiction!" and this may well be true, but we are considering a completely new kind of experiment, which involves the very first direct two-way contact between one atmosphered world and another. One cannot be too careful, and it will be only wise to keep a returned probe circling in Earth orbit while the most rigorous tests are carried out. I agree that the risks are slight, but they are not nil, and a single careless mistake might spell disaster. Probably the final tests will be carried out by scientists upon an orbiting space-station. Only when an immensely detailed analysis has been completed will the first samples from Mars be brought down to the Earth itself.

There is also the danger of carrying Earth contamination to Mars. It is quite likely that terrestrial bacteria could survive there; and if they spread, we would never be able to decide which organisms were indigenous to Mars and which were not – so that we would have lost our only chance of studying the planet in its mint condition, so to speak. Since several Russian probes have landed there already, the chance may be gone even now; but I think not, because every care was taken to make sure that the landed objects were completely sterile. No process can be completely foolproof, but it will indeed be bad luck if any bacteria have slipped through.

Later Vikings will carry "Martian rovers", or wheeled vehicles which are able to move around in much the way that the Soviet Lunokhods have done on the Moon, and undoubtedly the Russian plans allow for robots of the same kind. Note, too, that when a Viking arrives near Mars, and enters a closed orbit round the planet, it will be able to wait for well over a month if necessary before releasing its lander, to guard against the possibility of an all-obscuring dust-storm similar to that which greeted Mariner 9.

To sum up: Vikings 1 and 2 will land on Mars in 1976. We may expect a Rover-carrying probe in 1979 or thereabouts, and probably a returnable vehicle by 1982. The rest of the period up to 1990 is likely to be punctuated by further landers and "collectors", so that by the end of this time we should have a complete knowledge of what Mars is really like. This will be all-important in deciding whether it is worth sending a manned expedition there as soon as advances in rocketry permit. Certainly no astronauts can reach Mars in the chemical-propellant vehicles of today; nuclear power is essential.

Some time ago there was a reasonably well-supported theory according to which the Martian atmosphere contained poisonous oxides of nitrogen. This can now be ruled out; there is no reason to suppose that the atmosphere will be toxic, and neither is there much fear that the ground will be too soft or crumbly to support the weight of a space-craft. Moreover, the extremes of temperature are not so great as those on the Moon, and in every way Mars is much less unwelcoming than any other world within our range. Finally, there is always the fascinating prospect of finding life – of a sort; and if we do detect any trace of living organisms, we will probably have made the most far-reaching discovery in the history of science. Well before 1990, we ought to know.

6

TOURS~GRAND AND OTHERWISE: 1975 TO 1992

Space-ships present their designers with very special problems. The amount of propellant that can be carried is strictly limited, and so long as we have to depend upon chemical fuels there is never much to spare; moreover, it takes a very massive launcher to dispatch a very small probe toward its target world. There is also the awkward fact that sudden changes of course cannot be made, and the idea of taking evasive action to steer clear of, say, a wandering asteroid is absurd. Thirdly, journeys even to Venus or Mars are bound to take months; to reach Jupiter, beyond the asteroid belt, involves an outward trip lasting for over a year and a half, and the other giants are more remote still. Saturn's minimum distance from us is more than twice that of Jupiter, and when we consider Uranus or Neptune the times of travel become menacingly long.

Of course, nuclear fuels will provide at least a partial solution, because we will be able to use shorter flight-paths instead of depending mainly on an initial "boost" and using the Sun's gravitation to swing the probe into the right orbit to rendezvous with its target. However, nuclear rockets cannot be expected before the 1980s, and when planetary probes first came to the fore the theorists cast around to see whether Nature could be of any help.

By a lucky chance, Nature was – for once – co-operative. It was found that in the late 1970s the giant planets would be so arranged that the gravitational field of one could be used to swing a probe outward toward another. This led to the concept of the Grand Tour. Something of the sort has already been put into practice with Mariner 10, which by-passed Venus and used the gravitational field of that unprepossessing world to send it inward to an en-

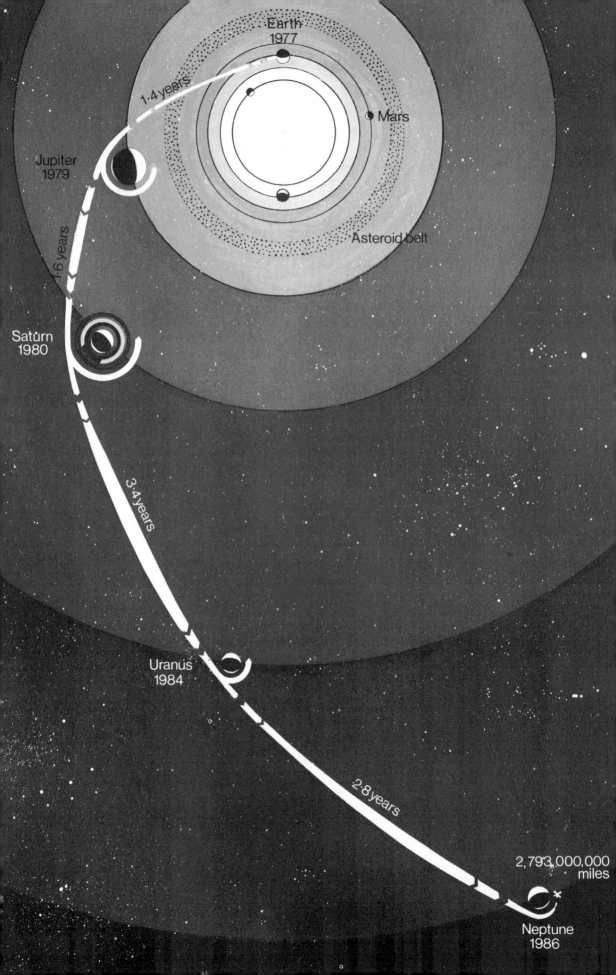

Earth
1977

1·4 years

Mars

Jupiter
1979

Asteroid belt

1·6 years

Saturn
1980

3·4 years

Uranus
1984

2·8 years

2,793,000,000
miles

Neptune
1986

counter with Mercury. But the Grand Tour project in its original form was much more ambitious, since it involved more planets and much longer journeys.

Jupiter is the key to the whole situation, because it is the senior member of the Sun's family, and is more massive than all the other planets combined. (Somebody once said that the Solar System is made up of the Sun, Jupiter, and assorted débris.) Jupiter has a mean orbital velocity of some 29,000 m.p.h., as against the 66,000 m.p.h. of the Earth. A probe approaching it would be accelerated by the tremendous gravitational pull, and both the direction and the velocity of the vehicle would be changed; although it would lose some of its new-found speed as it drew away from Jupiter, it would have picked up extra impetus. By the time that this had been lost, Saturn would be within range, so that the process could be repeated.

Everything depends on the relative positions of the giant planets, and the lining-up near the end of the 1970s will be more or less ideal; it will not recur for almost 180 years. One suggested Grand Tour is particularly intriguing. It runs as follows:–

1977 Sept. 4.   Launch from Earth.
1979 Jan. 28.   Encounter with Jupiter.   (Elapsed time, 1.4 years)
1980 Sept. 30. Encounter with Saturn.   (Elapsed time, 3   years)
1984 Jan. 2.   Encounter with Uranus.   (Elapsed time, 6.4 years)
1986 Nov. 8.   Encounter with Neptune. (Elapsed time, 9.2 years)

Without the "interplanetary billiards" technique, journeys to Saturn, Uranus and Neptune would take a great deal longer than this. The prospect was exciting, and elaborate plans were drawn

THE GRAND TOUR
One method of shortening the time of travel of a probe designed to rendezvous with the outer planets is to use the gravitational pulls of the various planets in turn. Thus an approach to Jupiter would be followed by an accelerated motion out toward Saturn, and so on. In the late 1970s and early 1980s the outer planets were suitably placed for this kind of procedure, and the Grand Tour was planned, as shown here. The original concept has been modified, largely for financial reasons, but the gravitational field of Jupiter will still be used to send probes out toward the more remote giants. The procedure has already been adopted for Pioneer 11.

up. There were several alternatives to the four-planet trip; one Tour would take in Jupiter, Uranus and Neptune, while another would cover Jupiter, Saturn, and the mysterious little world of Pluto.

One obvious difficulty was that a probe designed to wander about in the depths of the Solar System would have to be very reliable if it were to be of real use. Remember, too, that a radio command from Earth would take four hours to reach a vehicle moving in the vicinity of Neptune; and if anything went wrong the entire mission would have to be written off. I well remember talking to Dr. William Pickering, of the Jet Propulsion Laboratory at Pasadena, when I was over in California for a conference in 1969. "We have to design a complex space-craft which will continue to operate for ten years," he said. "And we don't have time to test it, because the flight must start less than ten years from now." He also commented: "If we want to go to three planets, I think 1979 to 1981 is the latest we can start."

All the same, Dr. Pickering and his colleagues were confident of success, and hoped to obtain information from Saturn or Uranus which would be just as detailed as that from the Mars probes which at that time had already been sent out (Mariners 6 and 7, which by-passed the Red Planet only a week or two after the first landing on the Moon by Armstrong and Aldrin). However, very powerful rockets were needed, and this was where Finance reared its ugly head. America was committed to the Apollo programme; there were also plans for space-stations and Shuttles, and the funds available were limited. At last the axe fell, and the chief victim was the original version of the Grand Tour.

Scientists in general were not pleased, and pointed out that we were wantonly missing a golden opportunity. Eventually there was something of a compromise. Pioneer probes to Jupiter were sanctioned, one to be launched in 1972 and the other in 1973, and there were at least possibilities of a probe which would be launched in 1977 to encounter first Jupiter and then Saturn. It was better than nothing, and Pioneer 10, the first of the Jupiter missions, was duly dispatched on schedule.

Jupiter itself is as unlike the Earth as it could possibly be. It has an equatorial diameter of over 88,000 miles, but its polar diameter is rather less, because its rapid rotation – less than ten hours – causes a flattening of the globe which is obvious when Jupiter is observed through even a small telescope. The disk is crossed by darkish belts and brighter zones, and there are various interesting features, notably the Great Red Spot. This is a huge oval object, sometimes 30,000 miles long and 9,000 broad, so that its surface area is greater than that of the Earth. Unlike the other spots, it seems to be at least semi-permanent, because it has been seen, on and off, ever since the seventeenth century. Sometimes it vanishes for a while, but it always returns.

It had long been established that Jupiter is made up of gas, at least in its outer layers, and that this gas is hydrogen-rich; there are quantities of methane and ammonia, both of which are hydrogen compounds, and it was assumed that there must also be some helium. Some authorities believed Jupiter to have a rocky core, overlaid by a deep layer of ice which was in turn overlaid by the hydrogen-rich atmosphere; according to another model, hydrogen was predominant all through the globe, though near the core the tremendous pressures would make the hydrogen behave in the manner of a metal. There was no serious belief that Jupiter could be a miniature sun, but it did seem likely that the temperature near the centre might amount to at least half a million degrees. As for the Great Red Spot – well, here too there were various theories. The Spot might be a semi-solid body floating among the Jovian clouds; it might be the top of a column of stagnant gas; it might be neither.

During the 1950s it had been found that Jupiter is a source of radio waves, and this implied the existence of a strong magnetic field, together with radiation zones much more intense than our own Van Allen belts. And quite apart from the planet itself, there were the satellites, of which four (Io, Europa, Ganymede and Callisto) were known to be large, though admittedly the other nine are cosmical dwarfs.

Altogether, Jupiter and its satellites were of special importance, and increased knowledge of them would clearly help toward a better understanding of the history and nature of the Solar System itself. It was hardly surprising that the progress of Pioneer 10 was followed with intense interest. The probe had to make its way through the asteroid belt; then it neared Jupiter, and in December 1973 the encounter took place.

There had been fears that the strong radiation would put the instruments on board the probe out of action. This is what very nearly happened. Pioneer went in to 81,000 miles from Jupiter, and the instruments were saturated; had the minimum distance been much less, they would have failed – and the mission would have failed with them. In the event, they survived, and sent a stream of information back to Earth. The magnetic field was indeed strong, but was "patchy", and not quite of the form that had been expected;

helium was positively identified in the Jovian atmosphere, and pictures were obtained, some of which showed the Red Spot splendidly.

There were also studies of two of the major satellites, Ganymede and Io. Of these members of Jupiter's family, Io is probably the most important, since it is known to have an effect upon the radio emission from the planet itself. It is slightly larger than our Moon, and Pioneer detected a slight ionosphere, though the pressure of the satellite's atmosphere was too low to be measured. Ganymede is larger still, though unfortunately some of the hoped-for results were lost because of an error in sending a command to the probe from Earth. However, this was no more than a minor disappointment; Pioneer emerged unscathed from its brief encounter, and went triumphantly on its way.

What next? So far as Pioneer 10 is concerned, nothing. It will never return; it will escape from the Solar System altogether, to become Man's first messenger to interstellar, as opposed to interplanetary, space. Contact with it will be maintained for some time, but eventually we will lose all track of it, and millions of years from now it may still be travelling outward between the stars, unseen and unheard. It carries a plaque, with a design which would – it is hoped – enable any civilization which finds it to identify its place of origin, but we cannot pretend that the chances of its being salvaged are very high!

So much for Pioneer 10. Even as it flew past Jupiter, its sister vehicle Pioneer 11 was on the way, and by the spring of 1974 was safely through the asteroid belt, which, incidentally, has been found to be much less "dusty" and therefore much less dangerous than had been expected. Pioneer 11 by-passed Jupiter on 5 December 1974, but the original plan was altered for two reasons. First, there was little point in merely repeating the Pioneer 10 mission, and a closer approach over the Jovian equator would mean certain failure of the instruments. Secondly, there was renewed interest in the second giant, Saturn, and it seemed reasonable to divert the probe there after it had fulfilled its main task.

When the decision was taken, in March 1974, Pioneer 11 was already at a vast distance from the Earth, and it says much for American technology that the whole sequence of events could be altered at so late a stage. On the new plan, an extra burst of power from the motors during the spring of 1974 altered the whole orbit, slowed down the space-craft, so that it passed within 26,000 miles of Jupiter at its closest approach – but this was over the polar region,

84

away from the worst effects of the radiation zones. It approached the planet from below Jupiter's south pole, swung up through the equatorial plane at an angle of 55 degrees, and left Jupiter well above the planet's north pole.

The close approach speeded Pioneer up to 110,000 m.p.h. relative to the planet, and this, plus the high angle of approach to the disk-shaped radiation belts, meant a very quick dash through the dangerous region. The stay in this danger-zone was reduced to a manageable level.

Even after this encounter, Pioneer 11's mission was not over, and it will attempt something even more ambitious: a rendezvous with Saturn. It will not, however, go "straight out". Again it must rely upon Jupiter's tremendous pull of gravity, and the resulting orbit for the probe will be anything but direct. Pioneer will fly in front of Jupiter (to the left, as seen from Earth) as it moves along. It will then pass behind the planet and emerge on the right-hand side, as shown in the diagram; it will have lost some of its velocity, and it will move into a looping orbit toward the opposite side of the Solar System. It will swing right across, and reach the neighbourhood of Saturn in September 1979.

The elapsed time between launch from Earth and encounter with Saturn will be $6\frac{1}{2}$ years. This is well beyond the originally-calculated lifetime for the instruments, but with luck the equipment will continue to function until Saturn has been by-passed. If so, the extra information will be sheer bonus; and Saturn, in its way, is every whit as fascinating as Jupiter. Through even a small telescope it is a magnificent sight, and it is unique in our experience.

The globe is basically of the same nature as that of Jupiter, though the overall density is lower, the belts are less conspicuous, and there is no Great Red Spot or anything comparable (though short-lived white spots do appear from time to time). So far there is no evidence of any magnetic field, but it would be rash to conclude that Saturn is non-magnetic, and neither can we rule out the possibility of radiation zones, though they are likely to be less powerful and therefore less dangerous to equipment than those round Jupiter. The beautiful ring-system is made up of countless particles whirling round Saturn in the manner of dwarf moonlets; recent radar measurements from Earth indicate that the particles are blocky rather than pebble-sized, and they may well be icy, or perhaps ice-coated. Though the system measures 170,000 miles from one side to the other, it is strangely thin, with a breadth which cannot be more than a few miles. When the ring-system is edge-on to the Earth, as happened in 1966 and will happen again in 1980, it cannot be seen at all except as a delicate line of light.*

It has been suggested that the rings represent the débris of an old satellite which ventured too close to the planet, and was literally torn to pieces. Whether this is true or not, Saturn still has a wealth of satellites left, and one of them – Titan – is proving to be of such interest that it is sometimes regarded as being just as important to theorists as Saturn itself.

Titan was discovered long ago, by Christiaan Huygens in 1655, and a small telescope will show it. Estimates of its diameter range between about 2600 miles and about 3500 miles, so that at any rate Titan is much larger than our Moon, and may be larger than the planet Mercury. What makes it unique is that it is the only satellite in the Solar System known to have a relatively thick atmosphere – and in fact the ground pressure of this atmosphere is of the order of 100 millibars, or ten times as great as the pressure at the surface of Mars. There may well be clouds there, even though the atmosphere is made up of methane and hydrogen and is utterly unbreathable by any life-form of the kind we know.

Yet the escape velocity of Titan is a mere 1.7 miles per second, less than that of the virtually airless Mercury. Titan's atmosphere is therefore of rather special type. There is a constant escape of gas, but this gas cannot break free from the pull of Saturn, and collects in a sort of doughnut-shaped ring round the planet, with Titan in the mid position. The result is that Titan re-collects the gas, and, to use an old cliché, what it loses on the roundabouts it gains on the swings. Probably the atmosphere is fully re-cycled every six years or so, which is a remarkable state of affairs.

We must also consider the effect of short-wave radiation upon

---

*It is incorrect to say, as some books do, that the rings vanish completely when edge-on to us. During the 1966 presentation I was able to follow them constantly, using the 10-inch refractor at the Armagh Observatory and my own 12½-inch reflector.

PIONEER 11'S ROUTE TO SATURN

Having by-passed Jupiter, in December 1974, Pioneer 11 was put into a path leading to a rendezvous with Saturn in 1979. This involved swinging round Jupiter and moving right across the Solar System. There is every hope that the equipment on board will still be working when Saturn is reached; but in any case a Mariner will rendezvous with the Ringed Planet not so very long after Pioneer 11 has done so.

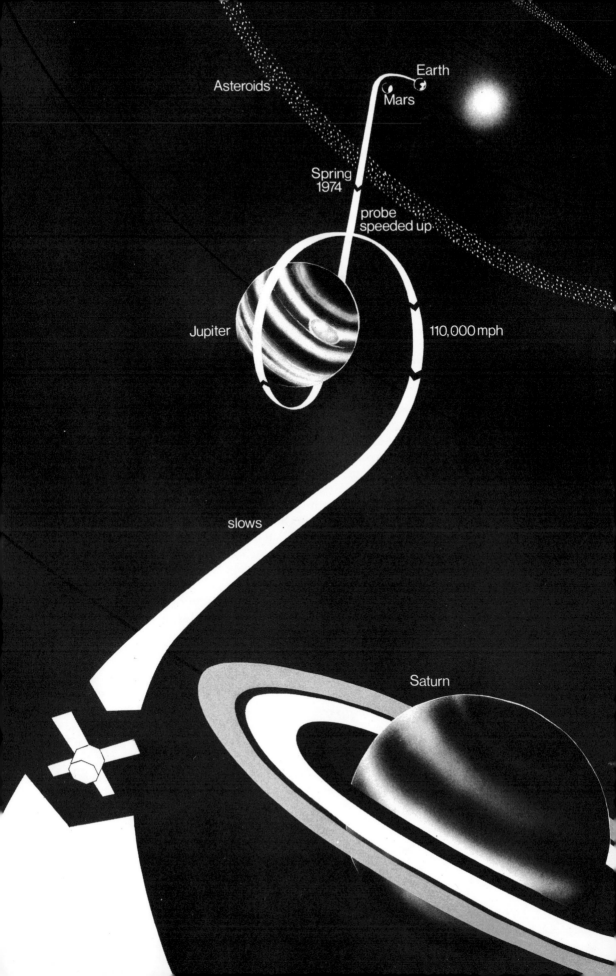

a primitive type of atmosphere such as Titan's. It has been suggested that hydrocarbons and amino acids may be produced; and amino acids would be of special significance, since they form a link in the chain of events leading to low-type biological activity. Of course, it would be most unwise to claim that there may be life on Titan. All the same, conditions there are so unusual that scientists are particularly anxious to study them. Certainly we cannot hope to find out much more except by using a space-probe, and Titan is now very high on the list of priorities. It could have many surprises in store for us.

Whether or not Pioneer 11 will survive as an operational unit until it encounters Saturn remains to be seen, but we must remember that it was not originally designed for any such mission, so that failure would not be too disappointing. More will be expected from the Jupiter–Saturn Mariner probe, which is scheduled for launching some time between mid-August and mid-September 1977. It will by-pass Jupiter in the summer of 1979, and will also survey the satellite system, with encounters with Io and also the outermost of the major moons, Callisto, which seems to be ice-covered.

The Mariner should reach the neighbourhood of Saturn in spring 1981, and will make fairly close approaches to two of the satellites: Iapetus, which is remarkable because of its variations in brilliancy – either it is irregular in shape, or has one bright and one darker hemisphere – and, of course, Titan. There will also be a near approach to the rings before the probe begins to recede, beginning its never-ending journey into interstellar space. By 1992 it should have reached a distance from the Sun equal to that of Neptune, though we can scarcely dare to hope that its transmitters will still be operating.

Even from Earth, Saturn is a glorious sight. What the Mariner pictures will show us we cannot yet tell, but they should be obtainable from about 80 days before the time of closest approach, and will presumably become more and more spectacular. The rings will be shown from angles which are unfamiliar to Earth-based observers, and the various shadow effects will be truly fascinating. Quite apart from this, we should at last learn something definite about the make-up of the ring system; we should decide whether or not Saturn has a magnetic field and radiation zones, and it is at least possible that Titan will be surveyed as clearly as Mars has been from the earlier Mariners, though there is always the chance that the relatively dense and cloudy Titanian atmosphere will hide the true surface.

88

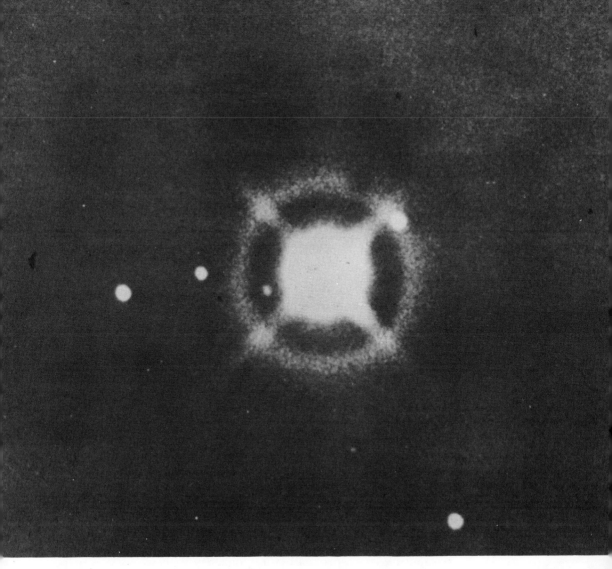

URANUS

Uranus has five satellites, all smaller than the Moon; they are shown on this photograph, taken by G. P. Kuiper with the 82 in. reflector at the Macdonald Observatory in America. Uranus itself is necessarily over-exposed.

Other projects are very much in the planners' minds. There may, for instance, be a Jupiter orbiter – that is to say, a probe which will approach the giant planet and then enter a closed path round it, as Mariner 9 did in the case of Mars; this would be of special value because of the variations in Jupiter's radio emission and magnetic field, the causes of which are still not known with any certainty. An orbiter may be achieved in 1981 or thereabouts. Next may come a probe designed to go right down on to – or, rather, into – Jupiter itself; inevitably it would be first silenced and then destroyed, but the data sent back during the last part of its active career would be of vital importance not only to astronomers but also to researchers in physics, chemistry and all allied branches of science.

Far beyond Saturn lie two more giants, Uranus and Neptune. Without the "billiards" technique, journeys to them would be painfully prolonged, but by using the pull of Jupiter it is feasible to send a vehicle to encounter Uranus by about 1985. And Uranus has mysteries of its own; it is much smaller than Jupiter or Saturn, but its diameter is almost 30,000 miles, and it may contain a great deal of methane, ammonia and water-vapour as well as hydrogen and helium. Unlike almost all other worlds, its axis is tilted at more than a right angle, so that the calendar there is particularly strange. There are 65,000 Uranian days in each Uranian year.

In all these long journeys there are extra problems to be faced. Solar cells cannot be used as sources of power, because the sunlight in the far reaches of the Solar System is too weak. The only solution is to use radio-isotope thermoelectric generators, which will have to function faultlessly over periods of years. Neither must we underestimate the difficult of sending the data back to Earth even when all the information has been collected by the probe itself. The return signal is astonishingly weak even for Jupiter, and will be much feebler still for Saturn and Uranus.

What of Neptune, last of the giants, and also the enigmatical little Pluto? Again we have tremendous increases in distance; the principles involved will be the same, with Jupiter once more playing a dominant rôle, but the difficulties are increased to an alarming extent. Yet I suspect that these remote worlds, too, will have been reached by space-probes before 1992. By then we may even be able to study the surface details of Pluto, the mysterious planet which, from Earth, appears as nothing more than a dot of light.

Uranus, which is dimly visible with the naked eye, was discovered by William Herschel in 1781. It is unfair to say that this achievement was a pure accident, because Herschel was busy upon a systematic "review of the heavens" and very little escaped him, but it is true that he was not looking for a new planet, and neither did he recognize it for what it was; his announcement to the Royal Society was headed "An Account of a Comet". The discovery of Neptune was made in a different way. Uranus persistently strayed from its predicted position; therefore, some unknown body must be tugging it out of place; and by studying the wanderings of Uranus, the position of the perturbing body was found. When Neptune was first identified telescopically, in 1846, it was as the result of a deliberate search in almost exactly the correct position.

Still things were not quite right. There remained some unexplained wanderings, and Percival Lowell, of Martian canal fame, repeated the procedure of tracking down an unknown planet. His task was even more difficult than the first had been, because the perturbations were smaller; all the same, his predicted planet – Pluto – was found, in 1930, almost exactly where Lowell had expected. Yet Pluto proved to be surprisingly small, and there were doubts as to whether it could genuinely cause the perturbations on other planets by which it had been tracked down.

Even today the problem remains unsolved. There may also be a tenth planet, far beyond either Neptune or Pluto, which is the real cause of the disturbances, and sporadic attempts have been made to calculate its position. Up to now the planet has not been located, and we are by no means certain that it exists at all. To carry out a full-scale hunt would occupy the time of a very powerful telescope for some years, because Planet Ten is bound to be very faint; it will look just like a star (as, indeed, Pluto also does), so that only its movement from one night to another will betray its nature. Since we have no real idea of its position, or even whether it is more than a myth, the prospect is rather daunting, and few scientists would agree in the idea of setting a major telescope aside for a long period in order to take part in a chase which might easily prove fruitless in the end.

This is quite understandable; but is there any other method? I suggest that there could be, but it will depend upon luck.

The first step is to decide upon the real size and mass of Pluto, and this is no easy task, because not even the Palomar 200-inch reflector will show a disk big enough to be measured accurately. The present diameter-estimate of 3700 miles must be regarded as highly unreliable. If it is correct, then Pluto comes midway in size between Mercury and Mars, but we do not really know. The best way to find out is to send a probe past Pluto. The movements of the vehicle would tell us Pluto's mass, and if pictures could be obtained we would learn the true diameter as well.

If Pluto does prove to be much more massive than we think, the real existence of Planet Ten may be in doubt; but I rather expect that Pluto is a genuinely small world.

After its encounter with Pluto, the probe will move outward – and if left to its own devices it will escape from the Solar System, as Pioneer 10 is in the process of doing and Pioneer 11 will do in the years following its pass of Saturn. But if a further burst of power could be applied, the probe could be put into a path which would

keep it in orbit round the Sun. Now, suppose that it remains in radio contact, so that its position is known, and that it comes within the sphere of influence of Planet Ten. We could then use the movements of the vehicle to tell us where Planet Ten is; and as soon as we have even an approximate position, we should be able to identify the planet optically.

I am well aware that this is highly speculative. Planet Ten may be non-existent. Even if not, it may be (indeed, probably will be) in the wrong position to affect the movement of a probe; a chance rendezvous in the vastness of space would be a piece of incredible good fortune. There is also the difficulty of keeping in touch with a vehicle which is moving at such an immense distance. But if the experiment is tried, if Planet Ten is convenient enough to be in the right area at the right time, and if technology can provide a probe which can be made to go on sending signals back to us for years after its main work is done – well, Planet Ten just *might* be located. I have not heard of this suggestion being made by anyone else, and I am quite ready for it to be greeted with derisive laughter. But to ape Edmond Halley:* If Planet Ten is eventually found by this method, I hope that posterity will not refuse to admit that the idea was put forward by an obscure English amateur in the 1970s!

Before leaving the subject of future automatic probes, we must not omit to take a brief look at some of the other bodies of the Solar System. For instance there are the asteroids, most of which keep to the region between the orbits of Mars and Jupiter; even the largest of them, Ceres, is no more than about 800 miles in diameter, and all of them must be totally devoid of atmosphere. Pioneers 10 and 11 have been through the asteroid zone, and we now know that there is much less dusty material there than had been thought, but the total number of asteroids must run into many thousands, even though most of the members of the swarm are mere chunks of

---

*In 1682 Edmond Halley, friend of Newton and later to become Astronomer Royal, observed a bright comet, and worked out that it must be the same as comets seen previously in 1531 and 1607. He predicted its return for 1758, and added that if this duly happened "posterity would not refuse to acknowledge" that this was discovered by an Englishman. The comet came back on schedule, and was discovered on Christmas Night, 1758, by a Saxon observer named Palitzsch. Since then Halley's Comet has returned in 1835 and 1910, and will be back once more in 1986. It is the only bright comet to have a period short enough for us to find out when to expect it.

DISCOVERY OF PLUTO.
Pluto is arrowed, and is seen to have moved compared with the stars: the upper photograph was taken three nights before the lower. The bright, over-exposed star is Delta Geminorum.

rock. (As I have commented, there is a good chance that Phobos and Deimos ranked as asteroids before being captured by Mars, and the minor members of Jupiter's satellite family, as well as Phœbe in Saturn's system, may be of the same type – though admittedly there are difficulties in the way of the capture hypothesis, and not all astronomers support it.)

It is hard to say when the first deliberate asteroid probe will be launched. However, Pioneers and Mariners to the outer giants must pass through the main zone, and the chance of an encounter with, say, Ceres would be rather too tempting to be missed, particularly if it ranked as sheer bonus. Equally interesting are those asteroids which swing away from the principal swarm, and may come close to the Earth. Such is Icarus, only a mile or two in diameter, whose minimum distance from us is a mere 4,000,000 miles. At perihelion it penetrates to the region within the orbit of Mercury, and must become red-hot, though its aphelion distance is nearly twice that of the Earth. Icarus has a "year" only slightly longer than ours, and a path which is highly inclined. At its last near approach, in 1968, it was followed for some time by radar, and a moderate telescope would show it as a speck of light.

I would hate to suggest a sojourn on Icarus, which, with its wild extremes of temperature, must be ranked as the Devil's Island of the Solar System. On the other hand an automatic beacon there would be of great value, and I expect that one will be dispatched eventually, though it will have to be designed with particular care and probably belongs to the period well after 1992.

Another exceptional asteroid is Eros, which is shaped like a sausage; it measures about 18 miles long by 9 wide. It was found as long ago as 1898, and proved to be very useful, because observations of its motions led to a good method of measuring the length of the astronomical unit or Earth–Sun distance. As this method is now obsolete there is no point in describing it here; but Eros was back again in early 1975, passing within 20 million miles of us. No plans for sending a probe to it have been announced, but at a future approach there could well be an attempt. Airless and lifeless though the asteroids undoubtedly are, they are of considerable importance to theorists who want to delve back into the past history of the Solar System, and it would be well worth taking a close look at one of these peculiar little worldlets.

Finally, what about the comets, those erratic wanderers which sometimes look much more imposing than they really are? A comet is not a solid, rocky body; it has been described as "the nearest approach to nothing which can still be anything", and it consists of small particles together with remarkably thin gas. Some comets move round the Sun in periods of a few years, so that we always know where and when to expect them, while others have orbits which are so eccentric that they return only once in many centuries. Such was Kohoutek's Comet, which was expected to make a brilliant display in late 1973 and early 1974, though it signally failed to do so. If you want to take another look at Kohoutek's Comet, I am afraid that you will have to wait for 75,000 years.

Obviously one can never land a probe on a comet – there is nothing solid enough on which to land – but it would be worth sending a vehicle through the gaseous tail, or even through the main comet itself. The chances of survival in the latter case would be rather low, because a collision with a particle of any real size would be disastrous, but the experiment would nevertheless be interesting, and tentative plans have been drawn up for contacting comets within the next decade. Encke's Comet, which comes back every 3.3 years, is one possible target. It is an old friend, and has now been seen at fifty different returns, the last being in 1974, but it is never a bright object.

Halley's Comet, due once more in 1986, is unco-operative inasmuch as it travels round the Sun in a retrograde or wrong-way direction, which causes problems at once. It may be necessary to send a probe out to a distance greater than that of the comet and then swing it inward, catching the comet up, so to speak, as it approaches the Sun. Once again we find that the procedure is not nearly so simple as might be thought.

Let us now sum up the immediate prospects. Pioneer 11 encountered Jupiter in late 1974 and then was sent on its way to Saturn. Viking is definitely scheduled to soft-land on Mars in 1976, and the Jupiter–Saturn Mariner ought to be launched in 1977. When we consider further Vikings, Jupiter orbiters and entry vehicles, probes to Titan, long-range missions to the outermost planets, asteroid encounters and comet probes, there are financial problems; all these missions have to compete for funds with the space-station and Shuttle projects, and there are tremendous technical difficulties as well. This certainly applies to the American programme, and presumably to the Russian as well. All the same, the prospects are exciting by any standards; and it will be a sad disappointment if, during the next fifteen years or so, we have not managed to send out messengers to the very limits of the Solar System.

7

THE LUNAR BASE: 1995 TO 2000

Question. "Do you think that in the foreseeable future it is going to be possible to set up scientific bases on the Moon, on a large scale?"

Answer. "I'm quite certain that we'll have such bases in our lifetime. Somewhat like the Antarctic stations and similar scientific outposts; continually manned, although there's certainly the problem of the environment, with the vacuum and the high and low temperatures of day and night. Still, in some ways it's more hospitable than the Antarctic. There are no storms, no snow, no high winds, no unpredictable weather; as for gravity – well, the Moon's a very pleasant kind of place to work in; better than the Earth, I think. It would be quite a pleasant place in which to do scientific work. It's quite practicable."

That conversation took place on 16 November 1970, and was heard by millions of people inasmuch as it was broadcast on BBC television during a "Sky at Night" programme. I was the questioner; the man who answered was none other than Neil Armstrong – the first pioneer to step out on to the surface of the Moon, and whose opinions are not to be lightly set aside!

Much has happened since that first trip. Armstrong and Aldrin spent only a brief period outside their grounded module, and were never out of sight of it. During the last Apollo, that of December 1972, Astronauts Cernan and Schmitt went for prolonged drives in their Lunar Rover, and explored a relatively wide area. It is fair to say that the three final missions were much the most important scientifically, and we know much more about the Moon now than we did when Neil Armstrong talked to me in Lime Grove. The

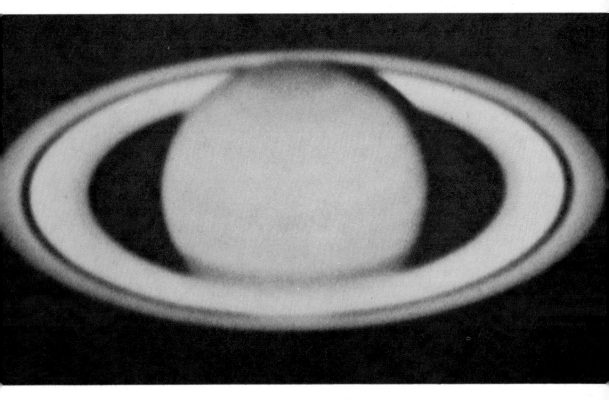

SATURN

The planet Saturn, photographed by G. P. Kuiper with the 82 in. reflector at the Macdonald Observatory, USA. The ring-system was wide open, so that the view was particularly beautiful.

SATURN FROM MARINER (*overleaf*)

In 1979 the Saturn Mariner will rendezvous with the planet, and we may hope for spectacular views of the ring system. It may have been preceded by Pioneer 11, but the Mariner has been specifically designed for studies of Saturn as well as Jupiter. Attention will also be paid to the largest of Saturn's satellites, Titan.

BENNETT'S COMET
This comet, discovered by the South African amateur astronomer
Jack Bennett, became a bright naked-eye object in 1970, and was the
most brilliant comet for many years. The photograph here was taken
by the discoverer.

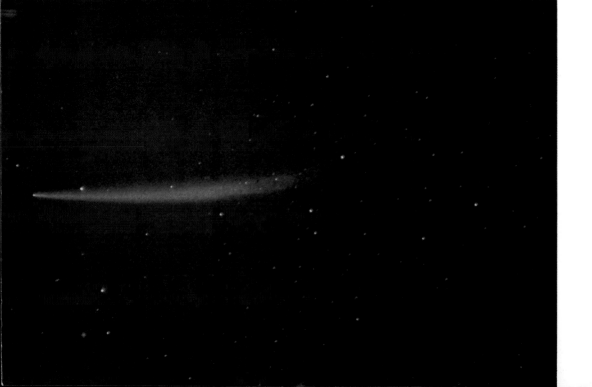

stage is set, and it would be a pessimist of the deepest dye who would question the logic of what he said.

Yet over-optimism is equally foolish, and let me repeat that if we are to establish Lunar Bases we must have vehicles of much greater adaptability and reliability than Apollo. So at the outset we must concede that everything depends upon the development of spacecraft which can go to and from the Moon in comparative ease and safety; and this brings us back to the Shuttle, because we cannot possibly separate one series of projects from all the rest.

Up to now I have been discussing programmes which have either been planned in detail or else are logical developments of them. When we come to discuss the future of man in space, there must inevitably be more in the way of speculation, but at least some vital facts have been firmly established. Skylab has proved that astronauts can survive for relatively long periods under conditions of zero gravity, so that there should be no danger in living on the Moon, where the gravitational pull is one-sixth of that to which we are accustomed on Earth. Secondly, there is nothing harmful on the Moon itself; there are no dangerous organisms, and no ground tremors strong enough to shake down any buildings erected there. Thirdly, there are no immense dust-drifts. The lunar surface is quite firm enough to support the weight not only of a space-craft, but also of a proper Lunar Base. And fourthly, the danger from meteorites falling on to the Moon and causing damage to any man-made structures there is very slight. Quite recently I re-read an excellent book which had been published in 1953, in which it was claimed that any Lunar Base would have to be built underground – because otherwise it would be soon battered to pieces by the steady cosmical bombardment. Of course, we cannot discount meteorite damage any more than we can on Earth; if the Siberian missile of 1908 had fallen upon a city instead of in frozen wasteland, the casualty list would have been horrifying. But meteorite falls likely to be harmful are probably very rare indeed, and it is significant that as yet no large fallen meteorites have been found on the Moon. One eminent astronomer–geologist, G. J. H. McCall, who – like myself – believes the main craters to be of internal origin, has even asked: "Where have all the meteorites gone?"

The science-fiction picture of a Lunar Base involves transparent inflatable domes, equipped with an efficient system of airlocks so that the pioneers can go in and out without losing any of the precious atmosphere. Ultimately this may not be far from the truth, but not, I think, as a start. It is even possible that the original Base will look rather more like a collection of tin cans. I do not propose to go deeply into the engineering problems, for the excellent reason that I am not an engineer; but we must remember that conditions on the Moon are totally alien, so that various structural materials will behave very differently there than they do here. We must also reckon with the tremendous temperature-range, from above +200 degrees Fahrenheit at noon on the equator down to below −250 degrees Fahrenheit during a night anywhere on the Moon.

As a start in my admittedly uncertain speculations, let us consider what the Russians may do – because there is always the chance that they will begin serious operations before 1990, which is probably the very earliest that the Americans will go back to the Moon.

Up to now the Soviet teams have concentrated upon robot exploration, and there is no doubt that their Lunokhods and recoverable Lunas have been technical triumphs. They have been less fortunate with their manned space research, though it is likely that they are fast overcoming the worst of their problems. When could they possibly start sending expeditions moonward? If they wanted to do so, they could begin by 1982 or thereabouts; and their first step might be to start dumping supplies at the selected site, which might well be one of the "mare" areas some way from the equator. The Mare Imbrium and the Mare Nubium are possibilities.

Supplies, unlike men, are expendable, and would provide excellent tests for the efficiency of the landing craft. A dozen or more vehicles would be sent, all coming down in the same area; they would carry supplies of all kinds, including power packs, elaborate radio equipment, and anything that might be needed in the way of constructional materials, together with medical kit. Once we find that all these are being dispatched, we may be fairly sure that Russian cosmonauts will follow.

The next step would be to send a reconnaissance crew. It is therefore quite on the cards that the first Russians on the Moon, unlike the first Americans, will not make a quick there-and-back trip, but will prepare for a more prolonged stay. They will start to get their future house in order, and upon their success will depend the rate at which the Base will develop. There will be continued reliance upon robot devices, and by that time there should also be several communications vehicles in orbit round the Moon.

It is impossible to guess at the design of the first Russian base. Certainly the planners will make it utilitarian rather than attractive! Massive rocket probes may be used as the initial "houses".

If so, they will land close together, and come to permanent resting-places, so that as soon as possible they can be linked together to provide roomier living quarters. In fact, the Base is likely to grow bit by bit, with each new section providing extra facilities. Within a few months of the first landing, we could see the Base in full operation. Its rôle would be to provide scientific information of all kinds; there will be laboratories as well as astronomical observatories, instruments for studying the nature of the Moon's globe, and medical centres in which insects and animals will both live and breed.

Obviously there must be provision for changes in personnel. It would be too much to expect any cosmonaut to stay on the Moon for a very long tour of duty, and this means that the Russians will not attempt any manned exploration until they have perfected reliable vehicles to take them to and fro.

This, as I have stressed, is guesswork on my part, and it is equally likely that the Russians will not try to set up any Base before the Americans are ready to start again toward the early 1990s. I am merely saying what, in my view, *might* happen. If it did, then American progress toward the Moon would be speeded up – not because of any political pique, but simply because the Russian experiences would provide an immense amount of detailed information, which would be available whether the Kremlin approved or not.

Much depends upon whether any useful materials are to be found on the Moon. The old, attractive idea of underground ice-sheets has been generally rejected (few practical lunar observers ever believed in it), and there is no hope of making an early Base self-supporting. Atmosphere, water and food must be taken from Earth, though later on it should be possible to cultivate plants inside the Base by the technique of what is usually termed hydroponic farming; the plants are suspended in nets above circulating liquids which provide the nutrients needed. This is by no means difficult. It works on the Earth, and there is no reason why it should not work just as well on the Moon.

Power, too, may be less of a problem than once anticipated, because for a continuous period of daylight, equal to two terrestrial weeks, the Sun will shine down from a cloudless sky. Solar power, plus nuclear power, should provide all that is needed, and the techniques for this could be worked out even now.

Next let us turn back to the American programme, which is by no means settled as yet, but which should be very much to the fore by 1990. Here too we have the prospect of an initial Base formed from the lunar modules which have taken the astronauts to the Moon, and it may be that the first expedition scheduled for a long stay will number about a dozen members. If this is achieved by – say – 1992, there ought to be a full-scale headquarters operating by 1995. And by 2000 we may see the first of the really elaborate Bases, possibly of the hemispherical form which was so widely favoured by the much-derided but often far-seeing fiction writers of the days before the Space Age began. One dome will act as the main living quarters; others will have their own specific rôles, and each will have its own system of airlocks for exit and entry. Clearly, it will be essential to make sure that if any dome develops a fault, the astronauts there can be satisfactorily accommodated elsewhere.

The lack of atmosphere, together with the reduced gravity, means that the research teams will be able to carry out experiments which are quite impossible on Earth. Moreover, the lunar surface is exposed to all the various radiations coming from space, and though this will admittedly be a mixed blessing it will be scientifically invaluable. As a radio and television relay station the Moon will also have a part to play; constant surveys of the Earth's weather systems will be possible, and as a site for an astronomical observatory the Lunar Base will be in a class of its own.

From Earth, optical astronomers are constantly plagued by the turbulence and dirtiness of the atmosphere, and the trouble increases according to the size of the telescope used. The Palomar 200-inch reflector works well, and we may expect much from the new Russian 236-inch, though it is true that the Soviet authorities are being rather reticent about it. But if it were possible to build, say, a 400-inch, the atmosphere would prevent it from being used to capacity, and might easily prevent it from being useful at all. On the Moon there is no atmosphere; and the lesser gravity would make for easier construction, both of optical surfaces and mechanical parts. Our view of the universe would be more than doubled in range.

More important still is the fact that from Earth we can study only a very limited part of the total electromagnetic range. We have the optical window, and there is also a radio window, but most of the radiations from beyond the Earth are blocked out, which is why we have had to use rocket vehicles for X-ray astronomy and other studies. This hazard, too, will not apply to the Moon. The prospects for radio astronomy are equally good, because the construction of very sizable equipment will be made easier by the fact that it will weigh less. Neither will there be serious interference from artificial transmissions, ranging from commercial broadcasting stations to unsuppressed motor-cycle engines.

Looking further ahead, the radio astronomer's ideal would be an observatory on the far side of the Moon, where shielding from Earth interference is complete. When this will be established must depend upon what happens during the next few decades, but with luck it should be in action before 2025. Contact with Earth, and with other Lunar Bases, will be maintained by using the communications satellites which will by then be orbiting the Moon.

These are some of the advantages. Alas, there are many disadvantages as well, and we may not yet appreciate them all.

A failure of some essential service in an Antarctic base is awkward, but not necessarily disastrous, because rescue can come in a matter of hours. This is not so with a base on the Moon, and it will not be so even when shuttle-type craft have been perfected. The greatest care must be taken not to over-populate the Moon. There will have to be enough space-craft available to ferry all the pioneers home in case of any real calamity, and this imposes a limitation at once. We can, I am sure, exclude any serious moon-quake, because the tremors are too slight to be harmful, and the really active period of the Moon's history ended many millions of years ago. I am thinking rather more of trouble caused by the long-term effects of radiation.

Skylab has disposed of the fear that astronauts will be harmed by radiations (or by lowered gravity) over periods of a few months, but as yet we do not know whether the effects will be cumulative. Without being unduly pessimistic, there is always a chance – mercifully slight – that some alarming symptoms will be suddenly found, affecting not one astronaut alone, but the whole colony. This might even call for prompt evacuation. I do not for a moment believe that it will happen, but we must be prepared for anything.

Also, it will be sensible to impose a strict limit on the tour of duty. Living under one-sixth g may be, and probably will be, not only harmless but positively comfortable – so long as one stays on the Moon. But what would happen to an astronaut who had spent, say, a couple of years there and then came home? On arrival he would feel heavier than lead; his muscles would have become so accustomed to "Moon weight" that they would be put under tremendous strain, and it would be some time before he could hope to readjust. If his stay on the Moon had been too long, he might be unable to readjust at all, and it is rather terrifying to think that a colonist who had spent several years on the rugged plains of the Mare Imbrium might find himself a permanent exile.

Unfortunately, there is no way of finding out except to try it.

Once the Base has been set up, tours of duty will be cautiously extended. There may or may not be a limit, but we do not yet know.

We must realize, too, that a man is not a robot. Neither is a woman, and this leads on to a problem which must already have been in the minds of long-term planners even though it has been too early to say much about it. Women can go into space; one (Valentina Tereshkova, now married to Cosmonaut Andriyan Nikolayev) has already done so. There is no reason why they should not go to the Moon, and I am quite sure that they will. A married man would not be willing to leave his wife and family for a pro-longed period, and a bachelor would be equally unlikely to join a lunar colony if it meant sacrificing any idea of marriage. And when men and women are together in an isolated Base, one does not need to be an Einstein (or a Spock) to realize that children will be born. There are unknown medical problems here. Whether a baby born on the Moon will be able to adapt to Earth conditions is yet another unknown factor, and it would be frightening to learn that a "Moon child" would have to spend his whole life on his adopted world. If so, of course, there would be a moral obligation to keep the Lunar Base in being, whatever the cost, and eventually there would be an entirely new branch of *homo sapiens*, possibly differing from ourselves not only in mental outlook but also in physical appearance. This is not likely to happen before 2025, because evolution is a slow process, but it could have happened before 3025.

In any case, our colonists must be able to lead lives which are as normal as possible, and this means that there must be ample provision for recreation; one cannot work all the time. There is no problem about what is termed "entertainment" (radio, television, etc.), and undoubtedly one of the early acts will be to establish a proper library. So far as exercise is concerned, there is no question but that regular activities will be essential – as, indeed, they are in space-stations of the Skylab or Salyut type; but it is tempting to go further, and decide what new sports will be worked out. As a cricketer, I can see endless possibilities. Not long ago I remember talking at a local cricket club dinner, and suggesting what could be done about limiting the size of the boundary, working out a new l.b.w. law, and bringing in regulations to help the luckless fast bowler whose run-up is bound to be seriously hampered! I look forward with interest to a *Lunar Sportsman's Manual* which must inevitably be published sooner or later. . . .

Returning to the more serious aspects, we must always remember that the team members of a Lunar Base will be in each other's company virtually all the time, and it would be idle to pretend that there will be no psychological problems. Initially the selection of astronauts, both male and female, will be just as rigorous as the selection of the Apollo and Skylab crews of today, but up to now we have been dealing with only a very few people, and a more populous Lunar Base is quite another matter. Moreover, if we look further ahead still to a colony of hundreds or thousands of people, the bad side of human nature means that crime cannot be ruled out, and there may even be a return to the tedious Earth "security checks" and the like. Fortunately, this is not something to worry about as yet, or before 2025.

With the international aspect, I admit to being more hopeful than some writers have been in the past. According to present trends, the original Bases will be either American or Russian; I hardly dare to hope that a combined team will be sanctioned before 1995, my tentative date for the setting-up of the first truly permanent Base. It is not likely that any other nation will be ready to go to the Moon on a major scale before then, though I suppose there must be certain reservations about China (of which my personal ignorance is complete). This means some emphasis on the national aspect, and on security, because even though the Moon is useless as a platform for launching offensive missiles it will be an excellent observing site from which no part of the Earth can be screened. Yet in the long run, the people who will matter on the Moon are the scientists – the physicists, chemists, engineers, biologists, medical men and astronomers; and in general these are the people who are anxious to collaborate with each other for the overall good of mankind. There is much more close collaboration in Antarctica, for instance, than some of the politicians like.

Governments and statesmen hold the purse-strings, and only they can rule whether or not the exploration of the Moon is to go on. Yet once it has started, it will be very difficult to stop, and collaboration will not be only desirable but also inevitable. Would a visitor from a Russian Lunar Base be unwelcome in the American headquarters, or vice versa? I hardly think so. Both are Earthmen.

This is bound to cause misgivings in both the White House and the Kremlin, to say nothing of Downing Street and in the office of whoever succeeds Chairman Mao. And yet, in my view, the politicians will be quite unable to put a stop to collaboration on the Moon – for the excellent reason that they will not be able to get there. Once such friendly collaboration has started, it will go on and will increase; and why should it not spread from the Moon to the Earth? I believe that the Lunar Bases will lead to a better understanding between the nations. Again I am reminded of the

Apollo 13 episode. When the astronauts were in grave danger, there was an all-too-brief period when the differences between East and West seemed to pale. It did not last; but a time may come.

One question which I am often asked is: "When will ordinary people – that is to say, non-technicians – be able to go to the Moon?" It is not a question which can be answered as yet, because it is both logical and necessary that the first and many succeeding expeditions will be made up of people who have been specially trained and have been sent to carry out definite tasks. But by 2000 it is possible that not every lunar traveller will be a scientist; and as the journeys become easier and easier, the training requirements will be relaxed. I do not suggest that before 2000 it will be feasible for members of the public to go to the Moon for a holiday, and I doubt whether this will become possible much before 2100, but one never knows.

From my own point of view, speaking as an observer who has always been particularly interested in the Moon, I must express a feeling of regret. By 1995 I will have reached the age of seventy-two, and I fear that I will never achieve my ambition of presenting a "Sky at Night" programme from the Mare Imbrium. Yet there is a chance that I may survive for long enough to see such a programme put on to the television screen by someone else, and this is something which even I would not have dared to suggest in April 1957, when I presented the first programme in the series and when the Space Age had not even begun.

8
THE MARTIAN BASE: 2020

There are no Martians – yet. There may be a certain amount of primitive life on the Red Planet; more probably there is none at all, but Mars is not a totally inert world. It has an atmosphere, albeit a very thin and unsatisfactory one, and it has both weather and seasons, while there is no positive proof that its giant volcanoes are extinct or even dormant. The more we learn about it, the more intriguing it seems to become, and there can be little doubt that it will be the first world to be reached beyond the Earth–Moon system.

Actually, much of what I have already said about the Lunar Base applies with equal force to any research station on Mars, because the atmosphere is quite useless for breathing and may be more or less useless as a screen against radiations. The ground pressure is nowhere more than 10 millibars, and over most of the planet it is less. This is equivalent to what we normally call a laboratory vacuum, and we must reluctantly abandon any ideas about "adapting ourselves" and strolling out during Martian mid-day with no extra protection other than an oxygen mask. Full pressure-suits will be needed all the time, and in this respect Mars is just as hostile as the Moon, so that astronauts who go there will always have to live under highly artificial conditions.

En passant, this seems the moment to refer briefly to the attractive idea of providing Mars with an atmosphere which would be of real value to us. I will not say much, because there is not the remotest prospect of our being able to do anything of the kind within the next fifty years, and I personally doubt whether it will be done within the next fifty centuries, if at all. In view of the near-vacuum conditions, we are starting practically from scratch; and even if we could somehow establish a dense, Earth-type atmosphere, the low escape velocity (3.1 miles per second) would mean constant leakage into space. The problem is even more difficult than that of altering an already-existing atmosphere, such as that round Venus. Still, there is always the chance that over a sufficiently long period Mars will perform the task for us, a point to which I will return later. Meantime, let us confine ourselves to the period between the present day and 2020 or thereabouts.

We know that chemically-powered rockets will not take men to Mars. A journey there would take months; add a similar time for coming home, plus a long delay on the planet before it and the Earth were suitably placed for the return trip, and the pioneers would be faced with a total absence of more than two years. The so-called minor problems, such as food and water supply, would become anything but minor. Remember, too, that an astronaut in orbit round the Earth, or even on the Moon, can cut his trip short and make haste for Earth if anything goes wrong, as in the case of Apollo 13. Once well away in interplanetary space, there can be no turning back until the mission has been completed.

No: we must have better means of travel, and things brings us back to the nuclear rockets which will presumably be developed during the 1980s. Once they have been thoroughly tested and proved to be both efficient and reliable, Mars will come within range. Until then, it will not.

Optimists claim that with steady, more or less trouble-free progress in research it should be technically possible to dispatch a Martian expedition by 1990; I will not demur; but I will nevertheless be very surprised if there are any attempts before the end of our century, and my own estimated date is around 2020. Technology is not the only consideration. Sending astronauts to Mars would be fantastically costly, and the planners will have to weigh up the pros and cons: what could a manned expedition achieve over and above what we can glean from automatic probes? Scientifically the returns would be great, but there is a limit to the amount of money which can be spent in securing a few samples of Martian dust, and the detractors would have a much better case than they had when trying to belittle the advantages of sending men to the Moon.

The main question, then, will be: Is there any hope of setting up a permanent base on Mars, and would it be worth while even if it could be done? I think that the answer, in each case, is "Yes", but if there is any doubt about it the first expeditions will be delayed well beyond the next fifty years. Moreover, we must not under-estimate the risks. The Americans did not attempt an Apollo landing on the Moon until they were sure that they had an over-whelmingly high prospect of success, and, as we know, their confidence was justified. It will take much longer to become equally confident about a trip to Mars.

Much will depend, too, upon what we learn about the planet during the next few years. As yet we cannot pretend to have enough information to guide us, despite the detailed studies from Mariner 9. It is not enough to know the atmospheric pressure, the surface temperature and similar basic facts. We have no idea, for instance, whether there is any sub-surface ice; if it exists, at least some of our worst problems will be reduced.

The Viking mission of 1976 may tilt the scales one way or the other, though we will still need to secure some samples and bring them back for analysis in one of the automatic recoverable probes which ought to be practicable well before 1990. Frankly, I have no faith in the theory that there may be harmful bacteria which will put Mars permanently "out of bounds", though I am the first to concede that we must make absolutely sure before drawing up final plans. Neither do I think there is any danger of finding a surface too fragile to support the weight of a space-craft or a building. Therefore, I am quite ready to look forward to the 2020 period, and speculate about the Martian Base which may be established.

A Base of some kind will be needed at once, because a protracted stay on Mars will have to be made even by the first pioneers, and to be cooped up in a small capsule for months would be out of the question. Both the physical and the mental strains would be in-tolerable. One must have room to move around.

Preliminary unmanned modules will be dispatched, and landed in the target area, wherever that may be. Presumably the site will be not too far from the equator, to take advantage of what warmth there is, and one possibility is the volcanic region of Elysium at around latitude 20°N., though of course there are many alterna-tives. Several modules will form the preliminary ingredients of a Base, and if they come down in an undamaged condition they can be air-filled and used by the astronauts who will follow; it is reassuring to have one's home waiting, so to speak. If a relatively large module of this kind is to be sent to Mars in one piece, it cannot be launched from the surface of the Earth. No doubt the modules will be assembled either in space or on the Moon, so that they will start their long journey from a vacuum environment. They will carry Martian Rovers of the same kind as those which carried the men of the last three Apollos across the surface of the Moon, and of course they will also take supplies of all kinds, including medical kit. The medical aspect is all-important; doctors and surgeons will be vital members of the expedition.

To complete the preliminary dumping process will take some time, and may last for several years. If started by 2015, all will be ready for the astronauts themselves by 2020.

As yet, when nuclear rockets are still in the stage of being designed, we have no real idea of how many crew-members can be taken on board one Martian craft. It is hardly likely that the first expedition will consist of one space-ship alone; there will be three, four or even more. We must allow for a minor fleet rather than solo vessel, which is yet another reason why the whole project will be so complicated and so expensive. If it is to be attempted at all, it must be international; no single nation could ever foot the bill.

It will indeed be a great moment when the first vehicle touches down. There is little room for error; the thinness of the atmosphere means that the lander cannot be provided with wings able to manœuvre it in the manner of an aeroplane, though the approaching vehicle will begin by entering a closed orbit round Mars and taking stock of the situation before the final descent. The interplanetary ship itself will be left in orbit, ready for the return journey, while the module carrying the astronauts will land in Apollo style, apart from the fact that parachutes will be used as well as rocket braking. Work must begin without delay, and there will be much to be done. The waiting modules must be checked and made ready for occupation; whether they can be linked together by a system of airlocks depends both upon their design and upon the accuracy of their positioning. Within a few Martian days of the arrival of the fleet, the colony should be in full operation.

Experience with unmanned probes has shown that there should be no trouble at all in keeping in contact with Earth, though things will be less easy when our world and Mars are on opposite sides of the Sun (that is to say, when Mars is in conjunction). Initially there will be only one Base, and this certainly takes us up to the end of the period between now and 2025. Later there will presumably be more, and travel between them will have to be either by ground transport or by rocket. The thin Martian atmosphere is useless so far as conventional aircraft are concerned; after all, not even our modern jet-liners can travel around at 100,000 feet, and at this height above the Earth the air-pressure is much greater than it is at the surface of Mars. I wonder, though, whether there is any chance of developing personal flying machines of any kind? The gravity is only one-third of that on Earth, so that the colonists will feel very light indeed. To be able to fly around would be an obvious advantage.

As with the Moon, so with Mars: the lack of useful atmosphere means that living conditions will be very artificial. Every effort must be made to conserve air and food. Water may or may not be a problem, depending upon whether there is any of it available in frozen form. The prospects are not too gloomy, because even though it is now thought that the white polar caps are made up chiefly of solid carbon dioxide there seems to be some water ice too, and there is also a little water vapour in the atmosphere, so that Mars is not completely arid.

I have said that the Lunar Base cannot be self-supporting in its early stages of development. The same could well be true of a base on Mars, and the situation is much trickier, because of the impossibility of ferrying extra supplies quickly and at short notice. By the time that a base is established on the Moon, a supply craft will be able to reach the lunar space-port within a day or so of a call for help, but even nuclear vessels will not allow for anything of the sort in the case of Mars. Therefore the colony must be as self-sufficient as possible, bearing in mind that it will have to become established during the very first expedition to the planet.

Mars has a rotation period of 24 hours 37 minutes, so that the day is only about half an hour longer than ours; the axial tilt is 24° as against our $23\frac{1}{2}$°, so that the seasons are of the same basic type even though they are much longer; and we can work out a Martian calendar easily enough. As with Earth, perihelion occurs during southern summer, so that the climate in the southern hemisphere is more extreme than that in the north. Reckoning in Martian days, we find the following lengths for the seasons:

Northern spring (southern autumn): 194 days.
Northern summer (southern winter): 177 days.
Northern autumn (southern spring): 142 days.
Northern winter (southern summer): 156 days.

making up a grand total of 670 Martian days in each Martian year. There could be twelve months, of which ten would have 56 days each and the remaining two 55, or there could be a system involving equal months and periodical Leap Years. No doubt a Martian World Calendar Society will eventually be formed to cope with the problem!

In one respect Mars differs from the Earth: there is no comparable moon. Both Phobos and Deimos are extremely small as well as being close-in, and they will be of little use as sources of illumination at night; Deimos in particular will look rather like a large, dim star. Phobos will cross the sky from west to east in a mere 4½ hours, during which time it will pass through more than half its cycle of phases fron new to full, but for long periods while it is above the horizon it will be eclipsed by the shadow of Mars. Deimos, whose revolution period is only about six hours longer than the rotation period of Mars itself, will stay up for 2½ days consecutively, and will go through its phase-cycle twice. Because both satellites move virtually in the plane of the equator they will be invisible from very high latitudes, but both will be on view from Elysium, where I have suggested that the first Base may be established.

I mention the dwarf moons here because there have been suggestions that they could be used as natural space-stations. It is not likely that they will be used as jumping-off points for Mars, but it is quite on the cards that communications stations will be set up on them at an early stage in the Martian venture. Of course, their gravitational pulls are weak; the escape velocity of Phobos is only 30 m.p.h., and that of Deimos even less. This is still too high for an astronaut to be able to leap clear by muscle-power alone, but anyone who jumped up from Phobos or Deimos would certainly rank as a temporary satellite of Mars. Incidentally, it is not impossible that modified space-guns could be used for sending payloads of the non-fragile variety from one of the satellites down to Mars.

Exploration will be a prime concern of the pioneers. Roving vehicles will venture out as far as possible; samples from different areas will be collected and analyzed; there will be instruments to measure the ground tremors which may well occur frequently; and there will be a major astronomical observatory. From Mars the Earth will appear as a bright inferior planet, bluish in colour, and showing phases just as Venus does to us. An unfamiliar feature of the night sky will be the numbers of naked-eye asteroids, since Mars is not too far from the inner edge of the asteroid zone.

This, then, may be the scene on Mars by 2020. The towering volcanoes, the swarms of craters, the deep rift valleys and the dry riverbeds will have been explored, at least over a limited region of the planet; a thriving colony will be in existence, and home on Earth volunteers will be queueing up to take part in this tremendous venture. Mars, like the Earth, will be alive.

And yet . . . remember, I say "*may be*". I am only too well aware of the immense technical difficulties, to say nothing of finance, and we cannot yet be confident that a full-scale expedition will be worth while in view of what can be done by using automatic rocket probes. I believe that Mars will tempt us; but I cannot be sure, and we will not know definitely for several years yet.

But suppose that the Base really does go ahead in the early part of the new century, and that the Martian colony becomes more and more populous. This is not impossible, though the problems to which I have referred in connection with the Lunar Base are magnified enormously. In any isolated colony there will be children, and we come back to the question of whether a baby born on another world will be able to adapt to conditions of Earth gravity. Experience from the Moon will have given better guidance before 2020, but the social, psychological and medical problems for Mars will present us with a completely novel situation, particularly as there may well be astronauts who volunteer to go to Mars and stay there permanently.

The 2020 Base will be no more than a start, but it will be of more significance in the long run than anything on the Moon. It will mark the beginning of a new race of men who will come to regard Mars, not the Earth, as their home planet. Looking well ahead, say to 2500, I can picture Martian libraries in which there are vivid descriptions of the early trips which will seem as remote as the Crusades are to us of 1975. There will be Martians who are proud of tracing their ancestry back directly to one of the pioneers, just as Americans are proud of being able to trace their family trees back to the Pilgrim Fathers.

We may even doubt whether the two worlds will be united under one Government and one code of laws. They may well have gone their separate ways, and the low gravity of Mars, together with the other differences in environment, may make a Martian as easily distinguishable from an Earthman as is a Pakistani from a Briton. This takes us far beyond the period of my present speculations – but if it does happen in the way I suggest, historians will agree that it all began around the year A.D. 2020.

9
MAN BEYOND MARS:
AFTER 2025

Strictly speaking, a book dealing with the next fifty years in space should be tactfully silent about ideas of sending men to any worlds other than the Moon and Mars. There are, of course, super-optimists who dream about exploring the system of Saturn and the gloomy, dimly-lit landscape of Pluto soon after the turn of the century, but I am not of their number. I have no doubt that all the solid-surfaced planets and satellites will be visited eventually, but certainly not in my lifetime, nor the lifetime of anyone now old enough to read this book.

Therefore I propose to be rather brief, and to do no more than indicate some of the eventual prospects.

Venus is actually the closest of the planets, and when at its nearest to us it is only about a hundred times as distant as the Moon. Against this, it is of peculiarly unpleasant nature. Quite apart from the high surface temperature and the crushing pressure, the atmosphere seems to be corrosive inasmuch as it contains sulphuric acid, and there is of course the tremendous abundance of smothering carbon dioxide. Little sunlight can penetrate that fuming, choking atmosphere. It has been said that Venus is remarkably like the conventional picture of hell, and I would be the last to disagree.

Suppose that an intrepid astronaut managed to land there in his space-ship, and then, with gay abandon, went through his airlock and stepped outside? His breathing apparatus might save him from being suffocated or poisoned, but he would certainly be both fried and squashed. This would be a pity, since, eerie though it might seem, the view of the landscape would be well worth seeing. The thick atmosphere would probably bend all rays of light so violently that the observer would appear to be standing in the bottom of a huge bowl, and it has even been claimed that an astronaut who looked straight ahead of him would see the back of his neck. This is an exaggeration, but certainly Venus must be classed as a psychedelic planet. Our outlook has changed drastically since the period before Mariner 2, when it was seriously proposed that there might be oceans at no more than a comfortably warm temperature, and that the prospects for colonizing Venus were rather better than for Mars!

I see no chance at all of men reaching Venus before 2025, or even 2125. Everything argues against it, and all exploration must be carried out by automatic vehicles, though even these have difficulty in surviving in this terrifyingly hostile environment. Indigenous life, as we know it, can be ruled out of court without further ado.

The existence of craters on Venus is also significant. They can hardly be of impact origin, since few meteorites could remain intact during a complete drop through so dense an atmosphere, and so the craters are probably volcanic. Erosion is presumably rapid and powerful, so that on the geological scale the craters should be young; and there are reasons for thinking that active vulcanism may be in progress even now, which will present an extra hazard. The Planet of Love is a world to be viewed from a respectful distance.

Oddly enough, Mercury may be reached before Venus, even though it is much more distant and is no friendlier than the Moon. Its thin atmosphere is absolutely useless for breathing or anything else, and the temperature range is extreme. On the other hand there are sound scientific reasons for wanting to send men to Mercury, because the planet would be an ideal site for a solar observatory which would presumably have to be serviced periodically – whereas the value of an expedition to Venus, even if it could be managed, is problematical. All the same, I do not believe that any astronaut will venture on to Mercury within the next fifty years.

Looking next at those parts of the Solar System which lie beyond the Earth's orbit, we come to the asteroids. Fiction writers have made great play of their potential mineral wealth, but I have grave reservations about this, and in any case the cost of "mining" an asteroid and bringing the material home would be prohibitive. A manned landing is not practicable as yet, or in the foreseeable future, and of course the arrival of a space-ship would be more of a rendezvous and docking than a true landing operation, because the gravitational pulls of even the major asteroids are negligible by everyday standards. A man on, say, as asteroid such as Flora would weigh less than a pencil does on Earth, so that he would be to all intents and purposes under conditions of zero gravity.

Robot stations on asteroids may be established before 2025, but that is as far as we can go. The idea of towing asteroids around and putting them into scientifically useful orbits has been mooted from time to time, but elementary calculations show that the power needed would be quite fantastic, and the whole scheme is likely to stay nothing more than a bright idea. Were I setting out to discuss the next five hundred years of space research, I would dwell longer among the asteroids; as things are, let us pass on.

With the giant planets, we come up against the problem of sheer distance. Even with nuclear rockets of the type which should be perfected by 2025, the journeys will still be depressingly protracted. This will not matter in the least for an automatic probe, but with manned space-ships the problems are alarming, particularly since astronauts travelling out beyond the orbit of Mars will be very much "on their own" without any hope of prompt rescue should things go wrong. And let us be realistic about one point: sooner or later there *will* be a tragedy in outer space. The meteoroid danger may be less than was once believed, but meteoroids are still plentiful enough, and eventually there is bound to be a collision between an interplanetary vessel and a lump of cosmical débris big enough to do serious harm.

Suppose, too, that one space-ship in a convoy heading out toward Jupiter develops trouble with its air purifiers? Transferring the members of a faulty ship over to a sound one reminds us of science-fiction stories involving the sharing out of what oxygen happens to be left. On this score alone I suggest that the first manned expedition out to the further reaches of the Solar System will be a gigantic operation involving not only crew-ships, but also several large, unpiloted supply vessels which will stay with the main fleet. To try anything on a smaller scale would be to court disaster.

There is not the slightest prospects of landing on any of the giant planets even when we have learned how to get there, because there is no visible solid surface, and an attempt to touch down on a layer of gas would be somewhat unwise. This means that we must concentrate upon the satellites of the giants, which are becoming more and more interesting as we collect information about them. Jupiter's major attendants are potential targets, particularly as observing sites for the planet itself, though we cannot yet tell how much a manned expedition would add to the results which could be transmitted home by automatic probes. Titan, in Saturn's system, is more important still, as the atmospheric pressure at its surface is ten times greater than on Mars, and the atmosphere appears to be cloud-laden.

Yet remember that Saturn is, on average, nearly 900 million miles from the Sun. Even allowing for space-craft which can go by shorter routes than those of today, and can travel at greater speeds, such a journey cannot be accomplished quickly, and the frailties of the human body have to be taken into account. Physically, a man (or a woman) might be able to endure a period of several years away from home, and of course "home" might mean the Moon or Mars rather than Earth, but there would be the mental strain to be considered as well, and elaborate tests will have to be carried out before the first trip is planned.

Frankly, I think we are looking well beyond 2025, and therefore it would be wrong for me to speculate further, but I do have one idea in my mind which may prove to be valid in the end. Let us assume that we develop permanent colonies on Mars, and that a new technology arises there, naturally based on that of Earth but with suitable modifications. The colonists will become adapted to the lower gravity, and so they and their descendants should be better able to endure conditions of weightlessness or reduced g. It could be that these pioneers, rather than men from Earth, will be the first to go to Callisto, Ganymede or Titan – and eventually, no doubt, to the satellites of Uranus, the senior moon of Neptune, and the bitterly-cold and mysterious Pluto. I would not like to hazard when this will be. It might be practicable by the end of the 21st century, or even earlier, but the task of carrying our explorations out into the depths of the Solar System may be left to the "new Martians".

10
BEYOND THE SOLAR SYSTEM: FROM 1975

It is difficult to send space-craft to the planets of our own Solar System; it is thousands or even millions of times more difficult to send them to the planets of other stars. We are still a long way from interstellar travel, and the chances of our achieving it before 2025 are absolutely nil. That being the case, the reader may ask why I am discussing it at all in the present book. The answer is that the exploration of the space between the stars has already begun, and so the foundations of all that lies ahead are to be found in our own time. At this moment Pioneer 10, the probe which by-passed Jupiter at the end of 1973, is heading outward – and it will never come back. It is our first messenger to escape from the Sun's influence, even though it seems condemned to travel aimlessly through the void until at last a collision with some cosmic particle destroys it.

Though we cannot yet dispatch probes out toward the stars with the slightest chance of keeping track of them, there is another way of exploring the universe. This, of course, is by telescope. The story began in the early seventeenth century with men such as Galileo, and it is still going on. Moreover, we are doing everything we can to search for traces of extraterrestrial life; and this seems to be a justifiable theme for me to follow up, because it will loom larger and larger in scientific thought between now and 2025.

I can dismiss the Solar System fairly quickly, because I have discussed the prospects earlier on. We may be quite sure that there is no life on the Moon, and never has been. We can also rule out the airless or near-airless worlds such as Mercury, all the asteroids, and all the satellites of the planets apart from Titan and possibly

Triton (the larger attendant of Neptune, which may have an atmosphere of reasonable thickness). Venus is really intolerable by any standards, and I for one have no faith whatsoever in the idea of alien life-forms floating gaily round inside the deep gas-layers of Jupiter or Saturn. We come back, as always, to Mars; and here there is still a chance that Viking and its successors will show up organic material, though I fear that the delightful Martian civilization so strongly championed by Lowell, less than sixty years ago, is just another myth.

I am again confining myself to life of the sort that we can understand, but it is hard to believe in the existence of "bug-eyed monsters" on Mars, Venus or anywhere else in our immediate vicinity, and I do not believe that they exist at all. Even so, the prospects for finding life elsewhere in the universe would be overwhelmingly strong if only we could find some means of communication, because civilizations of all kinds must surely exist.

To recapitulate: the Galaxy in which we live is made up of around 100,000 million stars, many of which are very like the Sun. (To take one example: the star Delta Pavonis, which is easily visible with the naked eye although it is too far south to be seen from Europe, lies at nineteen light-years from us, and is almost a carbon copy of the Sun.) The most powerful telescopes so far built are capable of photographing about a thousand million galaxies, and this does not give us the full score; there are others well beyond our range. In fact, the total number of stars which we know to exist is incredibly large, and to suppose that our own unimportant, typical Sun is unique in having a planet-family is absurd. It is also conceited, and we have learned from bitter experience that conceit never pays. There is a striking catalogue of humiliations, which may be summarized as follows:

1. *The Earth is flat, and quite unlike any other body in the universe.* – Disproved in Greek times.

2. *The Earth is the centre of the universe.* – Finally disproved during the "Copernican revolution", which began in 1543 with the publication of Copernicus' great book, and was ended by Newton in 1687.

3. *The Sun is an exceptionally important body.* – Disproved when it became clear that the stars are themselves suns. It is not possible to give a precise date, but if we say "the 18th century" we are not far wrong, even though the first star-distance was not reliably measured until 1838.

4. *The Sun lies near the centre of the Galaxy.* – Disproved early in our own century, when the late Harlow Shapley found that the

Sun, with its planet-family, lies well out toward the edge of the galactic system.

5. *Our Galaxy contains all the bodies in the universe.* – Disproved by Edwin Hubble in 1923, when he discovered that the so-called "starry nebulæ", such as the Andromeda Spiral, are far outside the confines of the Galaxy in which we live.

6. *Our Galaxy is exceptionally large.* – Disproved in 1952, when Walter Baade showed that all our previous distance-measurements of the external systems had been gross underestimates, and that the Andromeda Spiral, for instance, is about $1\frac{1}{2}$ times the size of our Galaxy as well as being more populous.

So the Earth is a typical planet; the Sun is a typical star; the Galaxy is a typical galaxy. Surely, then, no sane person can believe that the kind of life we have here is unique.

Though we have no positive proof of the way in which the planets were formed, modern theories are probably not very wide of the mark. I do not propose to discuss them in any detail, but it seems that the planets, including the Earth, were produced from material which was associated with the young Sun; the planets built up by accretion – that is to say, the clumping-together of the spread-out material. The process took a long time, but by 4500 million years ago the planets were already in existence as independent bodies moving in set paths round the Sun. What can happen to the Sun can happen to other stars also, and so why should not many of those stars we see at night-time have their own families?

This is, I must add, a reversal of earlier ideas current before the war. It used to be thought that the planets were literally torn off the Sun by the gravitational action of a passing star, and this would make the Solar System a true cosmical freak, because stellar encounters are rare. However, the whole "tidal theory" has long since been consigned to the astronomical scrap-heap.

Unfortunately, no telescope yet built will show a planet of another star. We have to depend upon less direct methods of investigation, and already there have been some notable results. Relatively nearby stars have individual or "proper" motions which are great enough to be measured with high accuracy; Barnard's Star, which lies at 6 light-years from us, travels across the sky so quickly that in a mere 180 years it covers a distance equal to the apparent diameter of the full moon. Now, a comparatively light-weight star associated with a massive planet will be pulled out of position, and will seem to weave its way across the sky instead of travelling in a regular line. The principle is quite straightforward, and was applied to some of the stars long before planet-families became a topic for discussion. Sirius, the brightest star in the sky, has a faint companion-star which was not discovered until 1862, but which had been predicted more than ten years earlier because of the irregular movement of Sirius itself.

The stars are not so very unequal in mass, despite their great range in size and luminosity. (Sirius is 26 times as powerful as the Sun, and has a diameter of over a million miles; its companion is only 1/10,000 as bright, and is smaller than the planet Uranus, but its mass is almost equal to that of the Sun, because it is amazingly dense.) A planet is much more elusive, and the perturbations produced on the parent-star are very slight indeed. However, very careful measurements carried out over periods of many years indicate that some of our stellar neighbours do indeed have planetary attendants; and Barnard's Star may have at least two. This is strong evidence in favour of the suggestion that Solar Systems are common in the universe. A solar-type star which is bereft of planets may be the exception rather than the rule.

Let us consider Barnard's Star in a little more detail. It is very faint by stellar standards, and is red; the two planets which have been indicated by the irregularities in its proper motion may be comparable in mass with Jupiter. Visually, they should be about magnitude +30, which is very dim indeed.* Even the Palomar reflector cannot photograph stars much below magnitude +23, and to observe the planets of Barnard's Star would mean using a much larger instrument. This could hardly be set up on Earth, because of the atmospheric turbulence. But what about the Moon?

We know that the airless Moon would be an ideal site for an observatory, and the low gravity would mean that some of the constructional problems would be eased. Well before 2025 there should be a telescope on the Moon with much more penetrating power than the Palomar 200-inch, and which should be capable of showing planets of other stars. This will be our first positive proof that the Sun's family is typical of many others. I may add that in my view, at least, a lunar observatory will be far more practicable than an observatory on a station in orbit round the Earth; there will be no stability problems – and because the Moon spins so slowly, the bodies in the lunar sky will seem to move at a majestic rate, which will help considerably with problems of telescope guiding.

---

*A star's magnitude is a measure of its apparent (not its true) brilliancy; the lower the magnitude, the brighter the star. Thus Aldebaran, in the Bull, is of magnitude 1; the Pole Star, 2; and so on. With the naked eye, stars of magnitude 6 are just visible on clear nights. The telescope in my observatory can take me down to magnitude +15. On the same scale, the Sun's magnitude is *minus* 26.

But it is not only with optical astronomy that we are concerned in our search for extraterrestrial life, and there may be a much greater chance of success by radio methods. Here, too, the Moon will be an invaluable site for an observatory, particularly when we can invade the hemisphere which is always turned away from the Earth and which is therefore blissfully shielded from artificial transmissions.

Radio astronomy began in the 1930s, with the fortuitous discovery of long-wavelength emissions from the Milky Way during some research by Karl Jansky in America. After the war, radio methods became of vital importance, and all kinds of sources were detected: the Sun, Jupiter, old supernovæ (the débris of past stellar explosions), and certain special galaxies which are still showing the results of tremendous outbursts inside them. Of course, it was never seriously suggested that these emissions were anything but natural, and the only temporary doubts came in 1969.

During that year, researchers at Cambridge University discovered some very peculiar, rapidly-varying radio sources which were absolutely regular and which seemed to be "ticking". There was a brief period of a few days when it was thought possible that we had at last picked up signals which were non-natural, and the Cambridge team wisely refrained from making any public announcement until they had satisfied themselves that this was not so. The "LGM" or Little Green Men theory was cast aside quickly, and we know that the sources are rapidly-rotating bodies, made up of the particles which we call neutrons. The bodies have been named pulsars, and apparently represent the final stages in the careers of once-powerful stars. The only pulsar which has been optically identified as yet lies in the Crab Nebula, a mass of expanding gas which is undoubtedly the wreck of a star which was seen to flare up in a final burst of glory in the year 1054. It is 6000 light-years away, so that it is hardly a near neighbour, but other pulsars must be more remote still.

I will return to the Cambridge reticence later. Meantime, what are the true chances of our picking up intelligent signals coming from space?

If we agree that there is no advanced life in the Solar System except (possibly!) on the Earth, we have to reckon in distances of light-years, remembering that a light-year – the distance covered by a ray of light in one year – is equal to almost six million million miles. All electromagnetic vibrations, including radio waves, move at the same speed as light itself: 186,000 miles per second. This means that our knowledge of the universe must always be out of date. For instance, we of 1975 are seeing Barnard's Star as it used to be in 1969, Sirius as it used to be in 1967, Polaris as it was in the time of the Crusades, Rigel in its Norman guise, and the Andromeda Spiral as it used to be before the last Ice Age began. The light from the most remote objects known to us started on its journey before the Earth existed as an independent body. Any messages, then, must come from the past. If we stretch the possibilities, and imagine that we will shortly pick up a signal from a planet circling a star in the Andromeda Spiral, this will not prove that intelligence exists there now; it will merely show us that intelligence existed in that part of the universe over two million years ago.

It is only during the past seventy years or so that radio has been developed on Earth, so that from the point of view of any alien civilization lying more than 70 light-years from us the Earth is still "radio quiet"; our transmissions have not yet penetrated further into the Galaxy. There are not many solar-type stars within this range, though a few do exist. The same reasoning can be applied in reverse; a civilization which began sending radio messages less than seventy years ago would still be untraceable unless it happened to lie within 70 light-years of us. Otherwise, the signals would still be on their way.

Seventy years is not much when we consider the vast time-scale of the universe, and it would be a strange coincidence if a civilization comparable with our own happened to develop in the same part of the universe *at the same time.* Moreover, we do not know how long a typical civilization lasts, and here we come back to problems which are more social than scientific. Ever since 1945 we have had the power to wipe out all life in Earth, and the nuclear weapons to do so are already stockpiled. There must have been many races in the Galaxy which reached this stage and duly destroyed themselves, leaving nothing behind but radiation-soaked death. One is tempted to wonder whether this is the norm – and whether *homo sapiens* will go the same way. If the average civilization lasts for only a few decades after its discovery of nuclear power, the chances of contact are considerably reduced.

All the same, intelligence must exist; and in the present state of our knowledge the only conceivable way to contact it is by radio. The first attempts were made in 1960, by scientists at Green Bank, West Virginia. Using the 85-ft. radio telescope there, they "listened out" at a wavelength of 21 centimetres in the hope of detecting signals which were rhythmical enough to be classed as artificial. This particular wavelength was selected for a very logical reason. The clouds of cold hydrogen spread through the Galaxy send out

125

radiations at 21 centimetres, and radio astronomers, wherever they may be, are bound to know about them; they will therefore concentrate upon this wavelength, and will reason that others are likely to do the same.

Project Ozma, as it was called, was limited in scope, because there could be no chance of receiving signals which could be decoded; but a rhythmical pattern would be immensely significant, and would at least be a start. Not surprisingly, nothing was found, and the experiment was soon halted, but since then the Russians have returned to something of the same kind – and there is every prospect of further experiments being made from research stations on the radio-quiet Moon well before 2025. The chances of success are slight, but not nil.

Project Ozma paid special attention to the two nearest stars which are reasonably like the Sun. Both are in the southern part of the sky; they are known as Tau Ceti and Epsilon Eridani, and both are rather more than 10 light-years from us. If, therefore, a radio astronomer there picked up the Ozma signals, we may expect a reply in the early 1980s; but I would not bank upon it!

The pulsar episode of 1969 has shown that it is unwise to jump to conclusions. At first examination the pulsars seemed too regular to be natural; it took some days to show that they were not artificial, and months to discover that they are neutron stars. But suppose that research teams based upon the Moon during the next fifty years or so did come across a signal which could not be explained in any way other than deliberate transmission from beyond the Solar System? The implications would be shattering, and one subject to be strongly affected would be religion.

There has always been an underlying feeling that our Earth is a special world, singled out for special treatment by a benevolent Deity. The Christian Church has Christ as its focal point; other religions have Buddha, Mahomed and so on, but in every case Man is given preferential treatment. For some years now scientists have been almost unanimous that there are other intelligences all over the universe, but the total lack of demonstrable proof has made the idea seem rather academic and remote. Some people refuse to accept it on purely ethical grounds, just as the Christian Church of the early seventeenth century stubbornly refused to believe in a moving Earth, and a few diehards of today still decline to believe in the theory of evolution.* Positive proof of the existence

*There is still an Evolution Protest Movement. See the chapter headed "Down with Darwin!" in my book *Can You Speak Venusian?*

TAU CETI
Tau Ceti, near the centre of this photograph, is 11 light-years away,
and is one of the nearest stars; it is rather smaller and cooler than the
Sun, but may well have a planetary system. Photograph by T. J. C. A.
Moseley (Armagh Observatory).

of another intelligent race, at a tremendous distance from us and
in a different planetary system, would be the final blow to our sense
of self-importance, and we would have to admit that the Earth is
merely one world in the eyes of God (whatever is meant by the
term "God"). There could be no logical case for a special relation-
ship, and the whole religious outlook would have to be drastically
amended "at a stroke".

I have a suspicion that the official Churches would decline to
accept the evidence until they were forced to do so, just as Roman
Catholicism, at least, is still distinctly wary of Darwin. I may be

127

wrong; I do not expect to live long enough to see my ideas put to the test, but certainly the religious upheaval would be more cataclysmic than anything which has happened during the past two thousand years. It might, in the long run, do us a great deal of good.

Obviously, there would be the closest possible investigation before declaring that a signal must be artificial. The only code which is universal in every sense of the word is that of mathematics. We did not invent mathematics; we merely discovered it, and there are all sorts of ways of transmitting information by mathematical symbols alone. The simplest would be a straight succession of numbers – in dot form, for instance. If we could pick up something of this kind:

```
          .

        .  .

      .   .   .

    .   .   .   .

  .   .   .   .   .
```

and so on, repeated time after time, it would be very difficult to find a natural explanation, and we would have to do no more than eliminate Earth-based transmission, either accidental or sent out as a deliberate hoax. More than that we could not do as a start.

There is always the chance that an advanced technology in another system would be able to pick up our ordinary broadcasts, but again the range is limited, because of the time needed for radio waves to spread out through space. I doubt whether the early broadcasts would get far in recognizable form, and of course most of our normal programmes are made at wavelengths which cannot penetrate the Earth's upper atmosphere at all. However, it is intriguing to speculate. Listeners on a planet moving round a star 35 light-years from us would not be receiving some of the broadcasts made by Mr. Churchill during the last war; pop music would have reached a planet at a distance of fifteen light-years or so (perish the thought!) but Neil Armstrong's "One small step" would not yet have reached any star beyond the three members of the Alpha Centauri system, plus Barnard's Star with its two possible planetary attendants.

Of course, I am putting purely imaginary and far-fetched cases, but I hope that the principle is clear. I will merely repeat that there is a very faint chance of our receiving some artificial signal from space in the foreseeable future, and the experiment is worth trying even though the prospects of success are far below one in a million. Sooner or later, if Man resists the temptation to blow up both himself and his homeland, contact must come.

The crippling limitation is the time taken for radio waves to travel through space, and this also applies, so far as we know, to any material object. I cannot believe that it will ever be feasible to send a manned space-ship out beyond the Solar System; my lack of faith in space-warps, time-warps, freezing techniques, and cosmical Noah's Arks is profound, though I am well aware that others do not agree. There remains the prospect of finding some method of communication which is not restricted in velocity, and we must not neglect the concept of instantaneous messages sent by the power of the brain alone.

I can imagine some readers giving an impatient grunt and throwing my book down at this point, assuming that I have passed beyond the bounds of credibility. I am not sure that they are right. We know very little about telepathy, but there is certainly "something in it", and there are many demonstrations on record which cannot be glossed over as due to faking or sheer luck. It is also worth remembering that one astronaut who has walked on the Moon (Mitchell, with Apollo 14) is firmly convinced of the truth of ESP or extra-sensory perception, and is now devoting his life to the study of it.

Telepathic research is still in its most rudimentary stage. We cannot even appreciate what its problems are, so that we have not the slightest idea of how to solve them; and there is a serious doubt as to whether our brains are sufficiently advanced for us to find out. Yet if I am right in saying that there are races in the Galaxy who have survived their testing-time and have developed to an extent beyond our comprehension, they may well be telepaths; and it is not inconceivable that they might be able to establish contact with our own primitive brains. Whether we would realize it is another matter, and there might be a danger that a recipient of an inter-stellar message would be promptly consigned to a lunatic asylum. There have been suggestions that this has happened already, though I do not for one moment believe so.

During the period between the present day and 2025, researches into telepathy and other "fringe sciences" will go on apace, and will be carried out with as much energy as astronautics. Since my own knowledge of the workings of the brain is very limited even by conventional standards, I am in no position to comment further, but I still maintain that there is the chance of a breakthrough at some time in the future. After all, what would Julius Cæsar have said if he had been told that within about two thousand years of his death it would be possible to sit in a house in Rome, look at a screen, and see and hear men walking about on the surface of the Moon? It would have been quite incomprehensible to him, and I suspect that we are much nearer to telepathic contact than Julius Cæsar was to television.

I must stop here, because it is most unlikely that the breakthrough, even if it comes, will be with us before 2025. Perhaps I have gone too far already. But I plead that it is the right time to speculate, because our thoughts have already moved out beyond the Solar System. Pioneer 10 carries its plaque; if any far-off race ever finds it, there will be no doubt of its artificial origin, and it is fascinating to picture some alien scientist pouring over the unfamiliar figures drawn by an unknown artist who will by then have been dead for many thousands of years.

11

THE MOUNTAIN TO MAHOMET:?

When I described the discovery of pulsars by Cambridge University scientists in 1969, I said that the decision to withhold any quick announcement was very wise indeed. The popular Press is not noted for its reticence. All sorts of wild stories have appeared from time to time, and anything backed up by the authority of Cambridge would have caused not only excitement but also, probably, alarm. Tradition, from the time of H. G. Wells onward, has made "other beings" both menacing and militant. It is less than forty years since a misleading broadcast of Wells' novel *The War of the Worlds* caused the famous panic in America.

If any truly artificial signal is received in the future, it is to be hoped that the discoverers will be as cautious as their Cambridge predecessors, so that nothing will be said until the facts have become clear – though, equally, it would be morally wrong to maintain secrecy for long. The effect of an actual visitation by an alien space-ship would be more devastating still, and no amount of security could keep it out of the public eye. Here, too, the chances of anything of the kind happening within the next fifty years are millions to one against; but – well, one never quite knows.

We have already gone through the flying saucer period, which was psychologically fascinating and which will be long remembered. Saucer (or, as they now prefer to be called, UFO) societies still exist; they add to the gaiety of nations, and they are quite harmless. Unfortunately the Selenites, Venusians and Martians have faded from scientific thought (if, indeed, they were ever there) and I cannot pretend to accept the findings of the Aetherius Society, whose members believe that their chairman, the Rev. Dr. George

King, is in telepathic touch with a Master of Light who lives on Venus and is allied with the Interplanetary Parliament which meets regularly on Saturn.* The idea that we have been contacted by space-ships from other Solar Systems is not much better, even if only because a denizen of a far-away planet would hardly want to undertake a journey of millions of millions of miles, spend a day or two looking casually around, and then depart once more. Remember, though, that civilization has existed on our world for only a very brief period, cosmically speaking. Had a visiting space-ship come here ten thousand years ago, he would have found an Ice Age planet with no technology whatsoever. Eighty million years ago, which again is not very long, he would have been greeted by nothing more intelligent than a dinosaur.

There is a distinction between saying that we *have not* been visited and claiming that we *cannot* be visited. We of 1975 can see no possible way of sending material objects through space at speeds greater than that of light – even in theory; in practice, of course, we cannot achieve even a ten-thousandth of this velocity. Older, alien technologies, which have left us far behind, may have found a means of doing so, and in this case there is no reason why they should not come here. This brings me to the point I made earlier about militant beings of the H. G. Wells variety. If every civilization has to survive a crisis period, after it has developed nuclear power but before it has reached a stage of enlightenment which precludes mutual destruction, all really advanced beings must be of the benevolent rather than the malevolent variety. If an alien craft suddenly landed in Hyde Park I, for one, would not be at all apprehensive about the intentions of its crew. The mission would surely be peace, not war.

On the other hand, what would be the reaction of the general public? Panic, I fear, and quite possibly a hasty attack which would sour relationships at once. There would be fear not only of any offensive action by the aliens, but also of possible contamination which could spread through the Earth's atmosphere and cause havoc. Much would depend on where the landing took place, though it is logical to assume that the visitors would have carried out preliminary surveys and decided upon the best way to approach us.

The situation would be complicated if the aliens were totally unlike ourselves in physical appearance. Though there is every

*Again, see my book *Can You Speak Venusian?* Members of the Aetherius Society are surely the pleasantest and the most astounding of all the Independent Thinkers.

STAR TRAILS
A time-exposure with a fixed camera will show star-trails, caused by
the Earth's rotation on its axis. This photograph, taken from Selsey
by Patrick Moore, shows the region of the bright northern
constellation of Cassiopeia.

TOUCHDOWN ON MARS (*overleaf*)
Artist's concept of a pioneer landing on the cratered surface of Mars,
with the beginnings of a Martian Base. Preparations will have to be
made for a reasonably prolonged stay even for the first expedition;
there can be no quick 'there and back' trip as for the Moon.

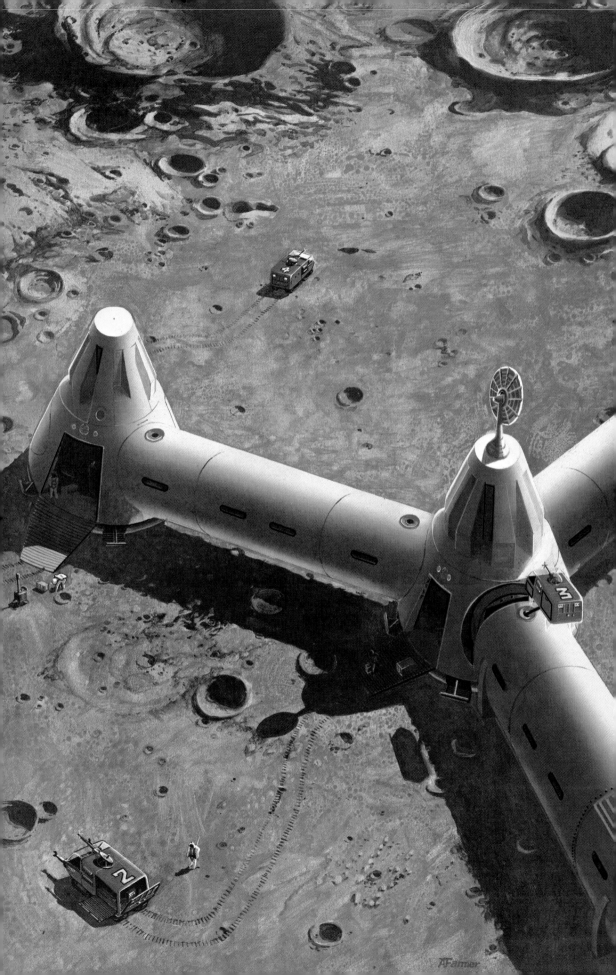

THE ANDROMEDA GALAXY, M.31
Messier 31, the Andromeda Spiral – the most famous external galaxy.
It is 2.2 million light-years from us, and is much larger than our own
Galaxy. Photographed with the 48 in. Schmidt telescope at Palomar;
reproduced by kind permission of the Hale Observatories.

THE 250 FT. JODRELL TELESCOPE (*below left*)
The great 250-foot radio telescope at Jodrell Bank – the 'dish' which
remains the world's most famous fully-steerable paraboloid, though
no longer the largest. Photograph by Patrick Moore, 1974.

THE CRAB NEBULA (*below right*)
Messier 1 (NGC 1952), the Crab Nebula in Taurus – the remnant of
the supernova of 1054. It is a source of radio waves, X-rays, etc., as
well as visible light, and is some 6000 light-years from us.
Photographed with the 200 in. reflector, and reproduced by kind
permission of the Hale Observatories.

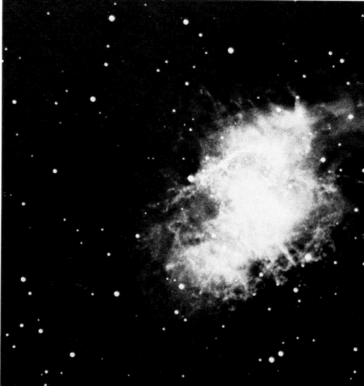

reason to think that intelligent life anywhere in the universe will be basically of our own type, and therefore not classifiable as a bug-eyed monster in the science-fiction meaning of the term, it is possible that a highly cultured alien coming here with the best of intentions would *look*, in our eyes, like a bug-eyed monster; and it is hard for us to break free from the prejudice that any advanced being must have one head, two arms, two legs and an upright stance.

I doubt whether communication would be a problem. It may sound absurd to suggest that an interstellar visitor would be able to speak English (or Russian, or Spanish) – but actually it is not unreasonable, because any technology capable of sending a star-ship to Earth would have learned our language by listening to our broadcasts. Also, the visitors would not want to risk a frightened and hostile reception. They might well decide to remain in orbit round the Earth and establish contact with us before making any physical landing.

The problems would not be over even if an understanding could be achieved quite quickly. An advanced civilization would be justified in regarding our world as very much of a shambles. There can have been few periods of late when there have not been at least a couple of wars going on somewhere or other, and nobody can seriously claim that any major nation is ruled in an intelligent and fair-minded way. Any alien capable of reaching us would be able to teach us a great deal, but everything would depend upon whether we are yet able to learn. On the whole, I rather doubt it.

This is an extra reason why I am certain in my own mind that contact will not be made yet. If by any remote chance we are under surveillance from outer space, the watchers will know that we are not ready for them; when we suggest that they might inter-vene in order to save us from ourselves, we are back to lurid science fiction; and the rational course would be to wait and let us work things out for ourselves. By alien standards the delay would not be long, because we are well and truly in the crisis period now. If we survive for another century or two, and provided that we solve the problem of ever-increasing population, we should be on the verge of real enlightenment. This will be the moment for contacts to be made. It is more likely that the mountain will come to Mahomet than that Mahomet will take the initiative and set out for the mountain.

By 2025 – no. By 3025 – well, perhaps.

SUMMARY

After these flights of fancy, let us return to the near-present and see what we can decide about the sequence of events during the coming half-century. Again I stress that quite apart from the purely scientific aspect there are all sorts of social and political complications to be taken into account. Any major war would through back the progress of space research for a very long time, perhaps for ever; a serious disaster, such as the failure of one of the next expeditions to the Moon, would cause a prolonged delay; there might be some kind of political reaction which would effectively cut off all funds, causing what we might call a period of interplanetary stagnation. Therefore, any forecasts must necessarily be arbitrary, and in any case it is hopeless to look ahead as far as 2025 and hope to be accurate.

Everyone has his (or her) own ideas, and I can do no more than give mine. For convenience, I have drawn up a table divided into three parts: the past, the planned, and the speculative.

(a) The Past

1957	Beginning of the Space Age, with Sputnik 1.
1959	First probes to the Moon: the Russian Luniks.
1961	First man in space: Yuri Gagarin.
1962	First successful planetary probe: Mariner 2 to Venus.
1962	First successful active communications satellite: Telstar.
1964	First really good close-range pictures of the Moon: Orbiter.
1965	First successful Mars probe: Mariner 4.

1966   First successful soft-landing on the Moon: Luna 9.
1968   First manned flight round the Moon: Apollo 8.
1969   Man on the Moon: Armstrong and Aldrin, in Apollo 11.
1971   First Mars orbiter: Mariner 9.
1973   First really successful space-station: Skylab.
1973   First probe to encounter Jupiter: Pioneer 10.
1974   First close-range pictures of Venus and Mercury: Mariner 10.
1974   Second encounter with Jupiter: Pioneer 11.
1975   First Russo-American space-docking. Apollo/Soyuz.

(b) The Planned

1976   First successful soft-landing on Mars: Viking.
1978   Landing of capsules on Venus (U.S.A.).
1978   Testing of the Shuttle.
1979   Further Viking to Mars, including a Mars Roving Vehicle.
1979   First encounter with Saturn: Pioneer 11.
1979   Encounter with Jupiter from a new Mariner vehicle.
1979   Polar orbiter round the Moon.
1981   Second encounter with Saturn, from the Mariner.

(c) Personal Speculation

1980   Probe to Encke's Comet.
1981   Orbiting vehicle around Jupiter.
1982   Probe to Mars capable of returning with samples.
1982   Venus orbiter.
1982   Mercury orbiter.
1982   First space-station intended to remain in orbit permanently.
1983   Russian network of automatic lunar stations now operative.
1985   Soft-landing of a transmitting probe on Mercury.
1985   First encounter with Uranus, from a Mariner.
1986   Probe to Halley's Comet.
1989   Encounters with Neptune and Pluto, from Mariners.
1990   Full operation of Earth satellite networks: meteorological, communications, etc. Several manned stations now in orbit.
1990   First Americans to go to the Moon since 1972.
1992   The first reasonably prolonged stay on the Moon.
1995   First Lunar Base.
2000   Major Lunar Base under international control.
2005   First asteroid beacons.
2015   "Dumping" of supplies and equipment on Mars.
2020   First Martian Base.

Dare I go further? Well: I would expect the lunar colony to be more or less of an independent entity by 2040, and the same would probably be true of the bases on Mars. Men might reach some of the satellites of Jupiter and Saturn by 2060, and by 2100 it should be possible to explore the whole of the Solar System.

By then, we may hope, a World Government on Earth will long since have been accepted, and there may even be a confederation of worlds, since the settlements on the Moon and Mars will have developed cultures of their own and many of the inhabitants will never have visited Earth – either because they have no wish to do so or because they cannot adapt to the much higher gravity.

I have made no provision for contact between our Solar System and another, either by radio, telepathy or physical travel, because I do not believe that this will happen within the next couple of hundred years, though it is not out of the question.

All this sounds decidedly futuristic, and it may prove to be so. If any visitors come here and find nothing but a scorched, un-inhabited planet we will have only ourselves to blame. We have the opportunity and the ability to prove ourselves, if only we can overcome our aggression.

There is one final point which nobody can question. If all goes well, and we really do progress as the centuries roll by, there will come a time when we will have explored the whole of our part of the universe; our whole outlook as a race of thinking beings will have been changed, and it will seem strange to look back over the years and reflect that there was a time when *homo sapiens* was confined to a single planet. The foundations are being laid now, and every-thing depends upon what we do in the immediate future. This is why the next fifty years are likely to be the most important in the whole history of mankind.

INDEX

142